Numerical Methods for Stochastic Computations

Numerical Methods for Stochastic Computations

A Spectral Method Approach

Dongbin Xiu

PRINCETON UNIVERSITY PRESS

PRINCETON AND OXFORD

Published by Princeton University Press, 41 William Street,
Princeton, New Jersey 08540

In the United Kingdom: Princeton University Press, 6 Oxford Street,
Woodstock, Oxfordshire OX20 1TW

press.princeton.edu

Library of Congress Cataloging-in-Publication Data
Xiu, Dongbin, 1971–
Numerical methods for stochastic computations : a spectral method
approach / Dongbin Xiu.
   p. cm.
Includes bibliographical references and index.
ISBN 978-0-691-14212-8 (cloth : alk. paper)
1. Stochastic differential equations—Numerical solutions.
2. Stochastic processes.   3. Spectral theory (Mathematics).
4. Approximation theory.   5. Probabilities.   I. Title.
QA274.23.X58 2010
519.2—dc22   2010014244

British Library Cataloging-in-Publication Data is available

This book has been composed in Times

Printed on acid-free paper. ∞

Typeset by S R Nova Pvt Ltd, Bangalore, India

Printed in the United States of America

10 9 8 7 6 5 4 3 2 1

To Yvette, our parents, and Isaac.

Contents

Preface   xi

Chapter 1   Introduction   1

  1.1   Stochastic Modeling and Uncertainty Quantification   1
    1.1.1   Burgers' Equation: An Illustrative Example   1
    1.1.2   Overview of Techniques   3
    1.1.3   Burgers' Equation Revisited   4
  1.2   Scope and Audience   5
  1.3   A Short Review of the Literature   6

Chapter 2   Basic Concepts of Probability Theory   9

  2.1   Random Variables   9
  2.2   Probability and Distribution   10
    2.2.1   Discrete Distribution   11
    2.2.2   Continuous Distribution   12
    2.2.3   Expectations and Moments   13
    2.2.4   Moment-Generating Function   14
    2.2.5   Random Number Generation   15
  2.3   Random Vectors   16
  2.4   Dependence and Conditional Expectation   18
  2.5   Stochastic Processes   20
  2.6   Modes of Convergence   22
  2.7   Central Limit Theorem   23

Chapter 3   Survey of Orthogonal Polynomials and Approximation Theory   25

  3.1   Orthogonal Polynomials   25
    3.1.1   Orthogonality Relations   25
    3.1.2   Three-Term Recurrence Relation   26
    3.1.3   Hypergeometric Series and the Askey Scheme   27
    3.1.4   Examples of Orthogonal Polynomials   28
  3.2   Fundamental Results of Polynomial Approximation   30
  3.3   Polynomial Projection   31
    3.3.1   Orthogonal Projection   31
    3.3.2   Spectral Convergence   33
    3.3.3   Gibbs Phenomenon   35

3.4	Polynomial Interpolation	36
	3.4.1 Existence	37
	3.4.2 Interpolation Error	38
3.5	Zeros of Orthogonal Polynomials and Quadrature	39
3.6	Discrete Projection	41

Chapter 4 Formulation of Stochastic Systems — 44

4.1	Input Parameterization: Random Parameters	44
	4.1.1 Gaussian Parameters	45
	4.1.2 Non-Gaussian Parameters	46
4.2	Input Parameterization: Random Processes and Dimension Reduction	47
	4.2.1 Karhunen-Loeve Expansion	47
	4.2.2 Gaussian Processes	50
	4.2.3 Non-Gaussian Processes	50
4.3	Formulation of Stochastic Systems	51
4.4	Traditional Numerical Methods	52
	4.4.1 Monte Carlo Sampling	53
	4.4.2 Moment Equation Approach	54
	4.4.3 Perturbation Method	55

Chapter 5 Generalized Polynomial Chaos — 57

5.1	Definition in Single Random Variables	57
	5.1.1 Strong Approximation	58
	5.1.2 Weak Approximation	60
5.2	Definition in Multiple Random Variables	64
5.3	Statistics	67

Chapter 6 Stochastic Galerkin Method — 68

6.1	General Procedure	68
6.2	Ordinary Differential Equations	69
6.3	Hyperbolic Equations	71
6.4	Diffusion Equations	74
6.5	Nonlinear Problems	76

Chapter 7 Stochastic Collocation Method — 78

7.1	Definition and General Procedure	78
7.2	Interpolation Approach	79
	7.2.1 Tensor Product Collocation	81
	7.2.2 Sparse Grid Collocation	82
7.3	Discrete Projection: Pseudospectral Approach	83
	7.3.1 Structured Nodes: Tensor and Sparse Tensor Constructions	85
	7.3.2 Nonstructured Nodes: Cubature	86
7.4	Discussion: Galerkin versus Collocation	87

Chapter 8   Miscellaneous Topics and Applications   89
  8.1   Random Domain Problem   89
  8.2   Bayesian Inverse Approach for Parameter Estimation   95
  8.3   Data Assimilation by the Ensemble Kalman Filter   99
           8.3.1   The Kalman Filter and the Ensemble Kalman Filter   100
           8.3.2   Error Bound of the EnKF   101
           8.3.3   Improved EnKF via gPC Methods   102

Appendix A   Some Important Orthogonal Polynomials in the Askey Scheme   105
  A.1   Continuous Polynomials   106
           A.1.1   Hermite Polynomial $H_n(x)$ and Gaussian Distribution   106
           A.1.2   Laguerre Polynomial $L_n^{(\alpha)}(x)$ and Gamma Distribution   106
           A.1.3   Jacobi Polynomial $P_n^{(\alpha,\beta)}(x)$ and Beta Distribution   107
  A.2   Discrete Polynomials   108
           A.2.1   Charlier Polynomial $C_n(x;a)$ and Poisson Distribution   108
           A.2.2   Krawtchouk Polynomial $K_n(x;p,N)$ and Binomial
                   Distribution   108
           A.2.3   Meixner Polynomial $M_n(x;\beta,c)$ and Negative
                   Binomial Distribution   109
           A.2.4   Hahn Polynomial $Q_n(x;\alpha,\beta,N)$ and Hypergeometric
                   Distribution   110

Appendix B   The Truncated Gaussian Model $G(\alpha,\beta)$ 113

References   117

Index   127

Preface

The field of stochastic computations, in the context of understanding the impact of uncertainty on simulation results, is relatively new. However, over the past few years, the field has undergone tremendous growth and rapid development. This was driven by the pressing need to conduct verification and validation (V&V) and uncertainty quantification (UQ) for practical systems and to produce predictions for physical systems with high fidelity. More and more researchers with diverse backgrounds, ranging from applied engineering to computer science to computational mathematics, are stepping into the field because of the relevance of stochastic computing to their own research. Consequently there is a growing need for an entry-level textbook focusing on the fundamental aspects of this kind of stochastic computation. And this is precisely what this book does.

This book is a result of several years of studying stochastic computation and the valuable experience of teaching the topic to a group of talented graduate students with diverse backgrounds at Purdue University. The purpose of this book is to present in a systematic and coherent way numerical strategies for uncertainty quantification and stochastic computing, with a focus on the methods based on generalized polynomial chaos (gPC) methodology. The gPC method, an extension of the classical polynomial chaos (PC) method developed by Roger Ghanem [45] in the 1990s, has become one of the most widely adopted methods, and in many cases arguably the only feasible method, for stochastic simulations of complex systems. This book intends to examine thoroughly the fundamental aspects of these methods and their connections to classical approximation theory and numerical analysis.

The goal of this book is to collect, in one volume, all the basic ingredients necessary for the understanding of stochastic methods based on gPC methodology. It is intended as an entry-level graduate text, covering the basic concepts from the computational mathematics point of view. This book is unique in the fact that it is the first book to present, in a thorough and systematic manner, the fundamentals of gPC-based numerical methods and their connections to classical numerical methods, particularly spectral methods. The book is designed as a one-semester teaching text. Therefore, the material is self-contained, compact, and focused only on the fundamentals. Furthermore, the book does not utilize difficult, complicated mathematics, such as measure theory in probability and Sobolev spaces in numerical analysis. The material is presented with a minimal amount of mathematical rigor so that it is accessible to researchers and students in engineering who are interested in learning and applying the methods. It is the author's hope that after going through this text, readers will feel comfortable with the basics of stochastic computation and go on to apply the methods to their own problems and pursue more advanced topics in this perpetually evolving field.

West Lafayette, Indiana, USA   *Dongbin Xiu*
March 2010

Numerical Methods for Stochastic Computations

Chapter One

Introduction

The goal of this chapter is to introduce the idea behind stochastic computing in the context of uncertainty quantification (UQ). Without using extensive discussions (of which there are many), we will use a simple example of a viscous Burgers' equation to illustrate the impact of input uncertainty on the behavior of a physical system and the need to incorporate uncertainty from the beginning of the simulation and not as an afterthought.

1.1 STOCHASTIC MODELING AND UNCERTAINTY QUANTIFICATION

Scientific computing has become the main tool in many fields for understanding the physics of complex systems when experimental studies can be lengthy, expensive, inflexible, and difficulty to repeat. The ultimate goal of numerical simulations is to predict physical events or the behaviors of engineered systems. To this end, extensive efforts have been devoted to the development of efficient algorithms whose numerical errors are under control and understood. This has been the primary goal of numerical analysis, which remains an active research branch. What has been considered much less in classical numerical analysis is understanding the impact of errors, or uncertainty, in data such as parameter values and initial and boundary conditions.

The goal of UQ is to investigate the impact of such errors in data and subsequently to provide more reliable predictions for practical problems. This topic has received an increasing amount of attention in past years, especially in the context of complex systems where mathematical models can serve only as simplified and reduced representations of the true physics. Although many models have been successful in revealing quantitative connections between predictions and observations, their usage is constrained by our ability to assign accurate numerical values to various parameters in the governing equations. Uncertainty represents such variability in data and is ubiquitous because of our incomplete knowledge of the underlying physics and/or inevitable measurement errors. Hence in order to fully understand simulation results and subsequently to predict the true physics, it is imperative to incorporate uncertainty from the beginning of the simulations and not as an afterthought.

1.1.1 Burgers' Equation: An Illustrative Example

Let us consider a viscous Burgers' equation,

$$\begin{cases} u_t + uu_x = \nu u_{xx}, & x \in (-1, 1), \\ u(-1) = 1, & u(1) = -1, \end{cases} \tag{1.1}$$

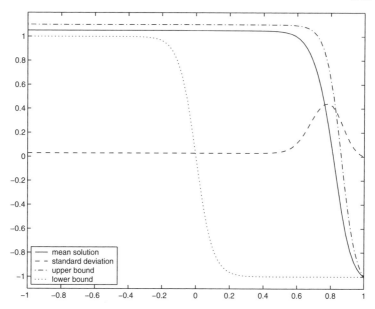

Figure 1.1 Stochastic solutions of Burgers' equation (1.1) with $u(-1, t) = 1 + \delta$, where δ is a uniformly distributed random variable in $(0, 0.1)$ and $\nu = 0.05$. The solid line is the average steady-state solution, with the dotted lines denoting the bounds of the random solutions. The dashed line is the standard deviation of the solution. (Details are in [123].)

where u is the solution field and $\nu > 0$ is the viscosity. This is a well-known nonlinear partial differential equation (PDE) for which extensive results exist. The presence of viscosity smooths out the shock discontinuity that would develop otherwise. Thus, the solution has a transition layer, which is a region of rapid variation and extends over a distance of $O(\nu)$ as $\nu \downarrow 0$. The location of the transition layer z, defined as the zero of the solution profile $u(t, z) = 0$, is at zero when the solution reaches steady state. If a small amount of (positive) uncertainty exists in the value of the left boundary condition (possibly due to some bias measurement or estimation errors), i.e., $u(-1) = 1 + \delta$, where $0 < \delta \ll 1$, then the location of the transition can change significantly. For example, if δ is a uniformly distributed random variable in the range of $(0, 0.1)$, then the average steady-state solution with $\nu = 0.05$ is the solid line in figure 1.1. It is clear that a small uncertainty of 10 percent can cause significant changes in the final steady-state solution whose average location is approximately at $z \approx 0.8$, resulting in a $O(1)$ difference from the solution with an idealized boundary condition containing no uncertainty. (Details of the computations can be found in [123].)

The Burgers' equation example demonstrates that for some problems, especially nonlinear ones, a small uncertainty in data may cause nonnegligible changes in the system output. Such changes cannot be captured by increasing resolution of the classical numerical algorithms if the uncertainty is not incorporated at the beginning of the computations.

1.1.2 Overview of Techniques

The importance of understanding uncertainty has been realized by many for a long time in disciplines such as civil engineering, hydrology, control, etc. Consequently many methods have been devised to tackle this issue. Because of the "uncertain" nature of the uncertainty, the most dominant approach is to treat data uncertainty as random variables or random processes and recast the original deterministic systems as stochastic systems.

We remark that these types of stochastic systems are different from classical stochastic differential equations (SDEs) where the random inputs are idealized processes such as Wiener processes, Poisson processes, etc., and tools such as stochastic calculus have been developed extensively and are still under active research. (See, for example, [36, 55, 57, 85].)

1.1.2.1 Monte Carlo– and Sampling-Based Methods

One of the most commonly used methods is Monte Carlo sampling (MCS) or one of its variants. In MCS, one generates (independent) realizations of random inputs based on their prescribed probability distribution. For each realization the data are fixed and the problem becomes deterministic. Upon solving the deterministic realizations of the problem, one collects an ensemble of solutions, i.e., realizations of the random solutions. From this ensemble, statistical information can be extracted, e.g., mean and variance. Although MCS is straightforward to apply as it only requires repetitive executions of deterministic simulations, typically a large number of executions are needed, for the solution statistics converge relatively slowly. For example, the mean value typically converges as $1/\sqrt{K}$, where K is the number of realizations (see, for example, [30]). The need for a large number of realizations for accurate results can incur an excessive computational burden, especially for systems that are already computationally intensive in their deterministic settings.

Techniques have been developed to accelerate convergence of the brute-force MCS, e.g., Latin hypercube sampling (cf. [74, 98]) and quasi Monte Carlo sampling (cf. [32, 79, 80]), to name a few. However, additional restrictions are posed based on the design of these methods, and their applicability is often limited.

1.1.2.2 Perturbation Methods

The most popular nonsampling methods were perturbation methods, where random fields are expanded via Taylor series around their mean and truncated at a certain order. Typically, at most second-order expansion is employed because the resulting system of equations becomes extremely complicated beyond the second order. This approach has been used extensively in various engineering fields [56, 71, 72]. An inherent limitation of perturbation methods is that the magnitude of the uncertainties, at both the inputs and outputs, cannot be too large (typically less than 10 percent), and the methods do not perform well otherwise.

1.1.2.3 Moment Equations

In this approach one attempts to compute the moments of the random solution *directly*. The unknowns are the moments of the solution, and their equations are derived by taking averages of the original stochastic governing equations. For example, the mean field is determined by the mean of the governing equations. The difficulty lies in the fact that the derivation of a moment almost always, except on some rare occasions, requires information about higher moments. This brings out the closure problem, which is often dealt with by utilizing some ad hoc arguments about the properties of the higher moments.

1.1.2.4 Operator-Based Methods

These kinds of approaches are based on manipulation of the stochastic operators in the governing equations. They include Neumann expansion, which expresses the inverse of the stochastic operator in a Neumann series [95, 131], and the weighted integral method [23, 24]. Similar to perturbation methods, these operator-based methods are also restricted to small uncertainties. Their applicability is often strongly dependent on the underlying operator and is typically limited to static problems.

1.1.2.5 Generalized Polynomial Chaos

A recently developed method, generalized polynomial chaos (gPC) [120], a generalization of classical polynomial chaos [45], has become one of the most widely used methods. With gPC, stochastic solutions are expressed as orthogonal polynomials of the input random parameters, and different types of orthogonal polynomials can be chosen to achieve better convergence. It is essentially a spectral representation in random space and exhibits fast convergence when the solution depends smoothly on the random parameters. gPC-based methods will be the focus of this book.

1.1.3 Burgers' Equation Revisited

Let us return to the viscous Burgers' example (1.1), with the same parameter settings that produced figure 1.1. Let us examine the location of the averaged transition layer and the standard deviation of the solution at this location as obtained by different methods. Table 1.1 shows the results by Monte Carlo simulations, and table 1.2 by a perturbation method at different orders. The converged solutions by gPC (up to three significant digits) are obtained by a fourth-order expansion and are tabulated for comparison. It can be seen that MCS achieves the same accuracy with $O(10^4)$ realizations. On the other hand, the computational cost of the fourth-order gPC is approximately equivalent to that for five deterministic simulations. The perturbation methods have a low computational cost similar to that of gPC. However, the accuracy of perturbation methods is much less desirable, as shown in table 1.2. In fact, by increasing the perturbation orders, no clear convergence can be observed. This is caused by the relatively large uncertainty at the output, which can be as high as 40 percent, even though the input uncertainty is small.

Table 1.1 Mean Location of the Transition Layer ($\bar{z}$) and Its Standard Deviation (σ_z) by Monte Carlo Simulations[a]

	$n = 100$	$n = 1,000$	$n = 2,000$	$n = 5,000$	$n = 10,000$	gPC
$\bar{z}$	0.819	0.814	0.815	0.814	0.814	0.814
σ_z	0.387	0.418	0.417	0.417	0.414	0.414

[a]n is the number of realizations, $\delta \sim U(0, 0.1)$, and $\nu = 0.05$. Also shown are the converged gPC solutions.

Table 1.2 Mean Location of the Transition Layer ($\bar{z}$) and Its Standard Deviation (σ_z) Obtained by Perturbation Methods[a]

	$k = 1$	$k = 2$	$k = 3$	$k = 4$	gPC
$\bar{z}$	0.823	0.824	0.824	0.824	0.814
σ_z	0.349	0.349	0.328	0.328	0.414

[a]k is the order of the perturbation expansion, $\delta \sim U(0, 0.1)$, and $\nu = 0.05$. Also shown are the converged gPC solutions.

This example demonstrates the accuracy and efficiency of the gPC method. It should be remarked that although gPC shows a significant advantage here, the conclusion cannot be trivially generalized to other problems, as the strength and the weakness of gPC, or any method for that matter, are problem-dependent.

1.2 SCOPE AND AUDIENCE

As a graduate-level text, this book focuses exclusively on the fundamental aspects of gPC-based numerical methods, with a detailed exposition of their formulations, basic properties, and connections to classical numerical methods. *No research topics* are discussed in this book. Although this leaves out many exciting new developments in stochastic computing, it helps to keep the book self-contained, compact, and more accessible to students who want to learn the basics. The material is also chosen and organized in such a way that the book can be finished in a one-semester course. Also, the book is not intended to contain a thorough and exhaustive literature review. References are limited to those that are more accessible to graduate students.

In chapter 2, we briefly review the basic concepts of probability theory. This is followed by a brief review of approximation theory in chapter 3. The material in these two chapters is kept at almost an absolute minimum, with only the very basic concepts included. The goal of these two chapters is to prepare students for the more advanced material in the following chapters. An interesting question is how much time the instructor should dedicate to these two chapters. Students taking the course usually have some background knowledge of either numerical analysis (which gives them some preparation in approximation theory) or probability theory (or statistics), but rarely do students have both. And a comprehensive coverage of both topics can easily consume a large portion of class time and leave no time for other material. From the author's personal teaching experience, it is better to go through probability theory rather fast, covering only the basic concepts and leaving other concepts as reading assignments. This is reflected in the writing of this book, as chapter 2 is quite concise. The approximation theory in chapter 3 deserves more time, as it is closely related to many concepts of gPC in the ensuing chapters.

In chapter 4, the procedure for formulating stochastic systems is presented, and an important step, parameterization of random inputs, is discussed in detail. A formal and systematic exposition of gPC is given in chapter 5, where some of the important properties of gPC expansion are presented. Two major numerical approaches, stochastic Galerkin and stochastic collocation, are covered in chapters 6 and 7, respectively. The algorithms are discussed in detail, along with some examples for better understanding. Again, only the basics of the algorithms are covered. More advanced aspects of the techniques, such as adaptive methods, are left as research topics.

The last chapter, chapter 8, is a slight deviation from the theme of the book because the content here is closer to research topics. The topics here, problems in random domain, inverse parameter estimation, and "correcting" simulation results using data, are important topics and have been studied extensively. The purpose of this chapter is to demonstrate the applicability of gPC methods to these problems and present unique and efficient algorithms constructed by using gPC. Nevertheless, this chapter is not required when teaching the course, and readers are advised to read it based on their own interests.

1.3 A SHORT REVIEW OF THE LITERATURE

Though the focus of this book is on the fundamentals of gPC-based numerical methods, it is worthwhile to present a concise review of the notable literature in this field. The goal is to give readers a general sense of what the active research directions are. Since the field is undergoing rapid development, by no means does this section serve as a comprehensive review. Only the notable and earlier work in each subfield will be mentioned. Readers, after learning the basics, should devote themselves to a more in-depth literature search.

The term *polynomial chaos* was coined by Nobert Wiener in 1938 in his work studying the decomposition of Gaussian stochastic processes [115]. This was long before the phenomenon of *chaos* in dynamical systems was known. In Wiener's work, Hermite polynomials serve as an orthogonal basis, and the validity of the approach was proved in [12]. Beyond the use of Hermite polynomials, the work on polynomial chaos referred in this book bears no other resemblance to Wiener's work. In the stochastic computations considered here, the problems we face involve some practical systems (usually described by partial differential equations) with random inputs. The random inputs are usually characterized by a set of random parameters. As a result, many of the elegant mathematical tools in classical stochastic analysis, e.g., stochastic calculus, are not directly applicable. And we need to design new algorithms that are suitable for such practical systems.

The original PC work was started by R. Ghanem and coworkers. Inspired by the theory of Wiener-Hermite polynomial chaos, Ghanem employed Hermite polynomials as an orthogonal basis to represent random processes and applied the technique to many practical engineering problems with success (cf. [41, 42, 43, 97]). An overview can be found in [45].

The use of Hermite polynomials, albeit mathematically sound, presents difficulties in some applications, particularly in terms of convergence and probability approximations for non-Gaussian problems [20, 86]. Consequently, generalized polynomial chaos was proposed in [120] to alleviate the difficulty. In gPC, different kinds of orthogonal polynomials are chosen as a basis depending on the probability distribution of the random inputs. Optimal convergence can be achieved by choosing the proper basis. In a series of papers, the strength of gPC is demonstrated for a variety of PDEs [119, 121].

The work on gPC was further generalized by not requiring the basis polynomials to be globally smooth. In fact, in principle any set of complete bases can be a viable choice. Such generalization includes the piecewise polynomial basis [8, 92], the wavelet basis [62, 63], and multielement gPC [110, 111].

Upon choosing a proper basis, a numerical technique is needed to solve the problem. The early work was mostly based on the Galerkin method, which minimizes the error of a finite-order gPC expansion by Galerkin projection. This is the *stochastic Galerkin* (SG) approach and has been applied since the early work on PC and proved to be effective. The Galerkin procedure usually results in a set of *coupled* deterministic equations and requires additional effort to solve. Also, the derivation of the resulting equations can be challenging when the governing stochastic equations take complicated forms.

Another numerical approach is the *stochastic collocation* (SC) method, where one repetitively executes an established deterministic code on a prescribed node in the random space defined by the random inputs. Upon completing the simulations, one conducts postprocessing to obtain the desired solution properties from the solution ensemble. The idea, primarily based on an old technique, the "deterministic sampling method," can be found in early works such as [78, 103]. These works mostly employed tensor products of one-dimensional nodes (e.g., Gauss quadrature). Although tensor product construction makes mathematical analysis more accessible (cf. [7]) the total number of nodes grows exponentially fast as the number of random parameters grows—the *curse of dimensionality*. Since each node requires a full-scale underlying deterministic simulation, the tensor product approach is practical only for low random dimensions, e.g., when the number of random parameters is less than 5.

More recently, there has been a surge of interest in the high-order stochastic collocation approach following [118]. A distinct feature of the work in [118] is the use of sparse grids from multivariate interpolation analysis. A *sparse grid* is a subset of the full tensor grid and can retain many of the accuracy properties of the tensor grid. It can significantly reduce the number of nodes in higher random dimensions while keeping high-order accuracy. Hence the sparse grid collocation method becomes a viable choice in practical simulations. Much more work has since followed, with most focusing on further reduction in the number of nodes (cf. [3, 75, 81, 104].

While most sparse grid collocation methods utilize interpolation theory, another practical collocation method is the *pseudospectral* approach, termed by [116]. This approach employs a discrete version of the gPC orthogonal projection operator and relies heavily on integration theory. One should keep in mind that in multidimensional spaces, especially for high dimensions, both interpolation and integration are challenging tasks.

The major challenge in stochastic computations is high dimensionality, i.e., how to deal with a large number of random variables. One approach to alleviate the computational cost is to use adaptivity. Current work includes adaptive choice of the polynomial basis [33, 107], adaptive element selection in multielement gPC [31, 110], and adaptive sparse grid collocation [35, 75, 104].

Applications of these numerical methods take a wide range, a manifestation of the relevance of stochastic simulation and uncertainty quantification. Here we mention some of the more representative (and published) work. It includes Burgers' equation [53, 123], fluid dynamics [58, 61, 64, 70, 121], flow-structure interactions [125], hyperbolic problems [17, 48, 68], material deformation [1, 2], natural convection [35], Bayesian analysis for inverse problems [76, 77, 112], multibody dynamics [89, 90], biological problems [37, 128], acoustic and electromagnetic scattering [15, 16, 126], multiscale computations [5, 94, 117, 124, 129], model construction and reduction [26, 40, 44], random domains with rough boundaries [14, 69, 102, 130], etc.

Chapter Two

Basic Concepts of Probability Theory

In this chapter we collect some basic facts needed for stochastic computations. Most parts of this chapter can be skipped, if one has some basic knowledge of probability theory and stochastic processes.

2.1 RANDOM VARIABLES

The outcome of an experiment, a game, or an event is random. A simple example is coin tossing: the possible outcomes, heads or tails, are not predictable in the sense that they appear according to a random mechanism that is too complex to be understood. A more complicated experiment is the stock market. There the random outcomes of brokers' activities, which in fact represent the economic environment, political interests, market sentiment, etc., are reflected by share prices and exchange rates.

The mathematical treatment of such kinds of random experiments requires that we assign a number to each random outcome. For example, when tossing a coin, we can assign 1 for heads and 0 for tails. Thus, we obtain a *random variable* $X = X(\omega) \in \{0, 1\}$, where ω belongs to the *outcome space* $\Omega = \{\text{heads, tails}\}$. In the example of the stock market, the value of a share price of a stock is already a random variable. The numbers $X(\omega)$ provide us with information about the experiment even if we do not know precisely what mechanism drives the experiment.

More precisely, let Ω be an abstract space containing all possible outcomes ω of the underlying experiment; then the random variable $X = X(\omega)$ is a real-valued function defined on Ω. Note here that for the abstract space Ω, it does not really matter what the ω are.

To study problems associated with the random variable X, one first collects relevant subsets of Ω, the *events*, in a class $\mathcal{F}$ called a σ-*field* or σ-*algebra*. In order for $\mathcal{F}$ to contain all relevant events, it is natural to include all the ω in the event space Ω and also the union, difference, and intersection of any events in $\mathcal{F}$, the set Ω, and its complement, the empty set $\emptyset$.

If we consider a share price X of a stock, not only the events $\{\omega : X(\omega) = c\}$ should belong to $\mathcal{F}$ but also

$$\{\omega : a < X(\omega) < b\}, \quad \{\omega : b < X(\omega)\}, \quad \{\omega : X(\omega) \le a\},$$

and many more events that can be relevant. And it is natural to require that elementary operations such as $\cap, \cup, {}^c$ on the events of $\mathcal{F}$ will not land outside the class $\mathcal{F}$. This is the intuitive meaning of a σ-field $\mathcal{F}$.

Definition 2.1. *A σ-field $\mathcal{F}$ (on Ω) is a collection of subsets of Ω satisfying the following conditions:*

- *It is not empty: $\emptyset \in \mathcal{F}$ and $\Omega \in \mathcal{F}$.*
- *If $A \in \mathcal{F}$, then $A^c \in \mathcal{F}$.*
- *If $A_1, A_2, \ldots, \in \mathcal{F}$, then*

$$\bigcup_{i=1}^{\infty} A_i \in \mathcal{F} \quad and \quad \bigcap_{i=1}^{\infty} A_i \in \mathcal{F}.$$

Example 2.2 (Some elementary σ-fields). The following collections of subsets of Ω are σ-fields:

$$\mathcal{F}_1 = \{\emptyset, \Omega\}$$
$$\mathcal{F}_2 = \{\emptyset, \Omega, A, A^c\}, \quad \text{where } A \neq \emptyset, A \neq \Omega,$$
$$\mathcal{F}_3 = 2^{\Omega} \triangleq \{A : A \subset \Omega\}.$$

$\mathcal{F}_1$ is the smallest σ-field on Ω, and $\mathcal{F}_3$ is the biggest one containing all possible subsets of Ω and is called the *power set* of Ω.

In practice, the power set is in general too big. One can prove that, for a given collection $\mathcal{C}$ of subsets of Ω, there exists a smallest σ-field $\sigma(\mathcal{C})$ on Ω containing $\mathcal{C}$. We call $\sigma(\mathcal{C})$ the *σ-field generated by $\mathcal{C}$*.

Example 2.3. In example 2.2, one can prove that

$$\mathcal{F}_1 = \sigma(\{\emptyset\}), \quad \mathcal{F}_2 = \sigma(\{A\}), \quad \mathcal{F}_3 = \sigma(\mathcal{F}_3).$$

2.2 PROBABILITY AND DISTRIBUTION

The concept of *probability* is used to measure the likelihood of the occurrence of certain events. For example, for the fair coin toss described in the previous section, we assign the probability 0.5 to both events, heads and tails. That is, $P(\{\omega : X(\omega) = 0\}) = P(\{\omega : X(\omega) = 1\}) = 0.5$. This assignment is based on empirical evidence: if we flip a fair coin a large number of times, we expect that about 50 percent of the outcomes will be heads and about 50 percent will be tails. In probability theory, the *law of large numbers* gives the theoretical justification for this empirical observation.

Some elementary properties of probability measures are easily summarized. For events $A, B \in \mathcal{F}$,

$$P(A \cup B) = P(A) + P(B) - P(A \cap B),$$

and if A and B are disjoint,

$$P(A \cup B) = P(A) + P(B).$$

Moreover,

$$P(A^c) = 1 - P(A), \quad P(\Omega) = 1, \quad P(\emptyset) = 0.$$

Definition 2.4 (Probability space). *A probability space is a triplet* $(\Omega, \mathcal{F}, P)$ *where* Ω *is a countable event space,* $\mathcal{F} \subset 2^{\Omega}$ *is the* σ-*field of* Ω, *and* P *is a probability measure such that*

1. $0 \leq P(A) \leq 1, \forall A \in \mathcal{F}$.
2. $P(\Omega) = 1$.
3. *For* $A_1, A_2, \ldots \in \mathcal{F}$ *and* $A_i \cap A_j = \emptyset, \forall i \neq j$,

$$P\left(\bigcup_{i=1}^{\infty} A_i\right) = \sum_{i=1}^{\infty} P(A_i).$$

Definition 2.5 (Distribution function). *The collection of the probabilities*

$$F_X(x) = P(X \leq x) = P(\{\omega : X(\omega) \leq x\}), \quad x \in \mathbb{R}, \tag{2.1}$$

is the distribution function F_X *of* X.

It yields the probability that X belongs to an interval $(a, b]$. That is,

$$P(\{\omega : a < X(\omega) \leq b\}) = F_X(b) - F_X(a), \quad a < b.$$

Moreover, we obtain the probability that X is equal to a number

$$P(X = x) = F_X(x) - \lim_{\epsilon \to 0} F_X(x - \epsilon).$$

With these probabilities we can approximate the probability of the event $\{\omega : X(\omega) \in B\}$ for very complicated subsets B of $\mathbb{R}$.

Definition 2.6 (Distribution). *The collection of the probabilities*

$$P_X(B) = P(X \in B) = P(\{\omega : X(\omega \in B\})$$

for suitable subsets $B \subset \mathbb{R}$ *is the distribution of* X.

The suitable subsets of $\mathbb{R}$ are called *Borel sets*. They are sets from $\mathcal{B} = \sigma(\{(a, b] : -\infty < a < b < \infty\})$, the *Borel* σ-*field*.

The distribution P_X and the distribution function F_X are equivalent notions in the sense that both of them can be used to calculate the probability of any event $\{X \in B\}$.

2.2.1 Discrete Distribution

A distribution function can have jumps. That is,

$$F_X(x) = \sum_{k:x_k \leq x} p_k, \quad x \in \mathbb{R}, \tag{2.2}$$

where

$$0 \leq p_k \leq 1, \quad \forall k, \quad \sum_{k=1}^{\infty} p_k = 1.$$

The distribution function (2.2) and the corresponding distribution are *discrete*. A random variable with such a distribution function is a *discrete random variable*.

A discrete random variable assumes only a finite or countably infinite number of values $x_1, x_2, \ldots$ and with probability $p_k = P(X = x_k)$.

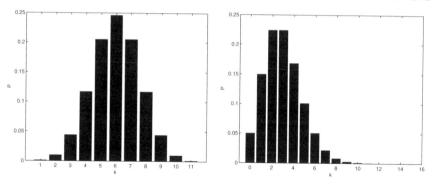

Figure 2.1 Probability distribution functions. Left: binomial distribution with $n = 10$, $p = 0.5$. Right: Poisson distribution with $\lambda = 3$.

Example 2.7 (Two important discrete distributions). Important discrete distributions include the *binomial distribution* $B(n, p)$ with parameters $n \in \mathbb{N}_0 = \{0, 1, \dots\}$ and $p \in (0, 1)$:

$$P(X = k) = \binom{n}{k} p^k (1 - p)^{n-k}, \qquad k = 0, 1, \dots, n,$$

and the *Poisson distribution* $P(\lambda)$ with parameter $\lambda > 0$:

$$P(X = k) = e^{-\lambda} \frac{\lambda^k}{k!}, \qquad k = 0, 1, \dots.$$

Graphical illustrations of these two probability distributions are shown in figure 2.1.

2.2.2 Continuous Distribution

In contrast to discrete distributions and random variables, the distribution function of a *continuous random variable* does not have jumps; hence

$$P(X = x) = 0, \qquad \forall x \in \mathbb{R},$$

or, equivalently,

$$\lim_{\epsilon \to 0} F_X(x + \epsilon) = F_X(x), \qquad \forall x; \qquad (2.3)$$

i.e., a continuous random variable assumes any particular value with probability 0. Most continuous distributions have a *density* f_X:

$$F_X(x) = \int_{-\infty}^{x} f_X(y) dy, \qquad x \in \mathbb{R}, \qquad (2.4)$$

where

$$f_X(x) \geq 0, \quad \forall x \in \mathbb{R}, \qquad \int_{-\infty}^{\infty} f_X(y) dy = 1.$$

Example 2.8 (Normal and uniform distributions). An important continuous distribution is the *normal* or *Gaussian* distribution $\mathcal{N}(\mu, \sigma^2)$ with parameters $\mu \in \mathbb{R}$,

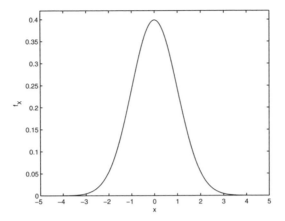

Figure 2.2 Normal distribution function with $\mu = 0, \sigma^2 = 1$.

$\sigma^2 > 0$. Its density is

$$f_X(x) = \frac{1}{\sqrt{2\pi\sigma^2}} \exp\left[-\frac{(x-\mu)^2}{2\sigma^2}\right], \qquad x \in \mathbb{R}. \tag{2.5}$$

The density of $\mathcal{N}(0, 1)$ is shown in figure 2.2.

The *uniform* distribution $U(a, b)$ on (a, b) has density

$$f_X(x) = \begin{cases} \frac{1}{b-a}, & x \in (a, b), \\ 0, & \text{otherwise,} \end{cases}$$

which is a constant inside (a, b).

2.2.3 Expectations and Moments

Important characteristics of a random variable X include its expectation, variance, and moments. The *expectation* or *mean value* of a random variable X with density f_X is

$$\mu_X = \mathbb{E}[X] = \int_{-\infty}^{\infty} x f_X(x)dx.$$

The *variance* of X is defined as

$$\sigma_X^2 = \text{var}(X) = \int_{-\infty}^{\infty} (x - \mu_X)^2 f_X(x)dx.$$

The mth *moment* of X for $m \in \mathbb{N}$ is

$$\mathbb{E}[X^m] = \int_{-\infty}^{\infty} x^m f_X(x)dx.$$

For a real-valued function g, the expectation of $g(X)$ is

$$\mathbb{E}[g(X)] = \int_{-\infty}^{\infty} g(x) f_X(x)dx.$$

Similarly, for a discrete random variable X with probabilities $p_k = P(X = x_k)$, we have

$$\mu_X = \mathbb{E}[X] = \sum_{k=1}^{\infty} x_k p_k,$$

$$\sigma_X^2 = \text{var}(X) = \sum_{k=1}^{\infty} (x_k - \mu_X)^2 p_k,$$

$$\mathbb{E}[X^m] = \sum_{k=1}^{\infty} x_k^m p_k,$$

$$\mathbb{E}[g(X)] = \sum_{k=1}^{\infty} g(x_k) p_k.$$

Often we study $\mathbb{E}\left[(X - \mu_X)^m\right]$, the *centered moments*. The expectation or mean μ_X, is often regarded as the "center" of the random variable X or the most likely value of X. The variance σ_X^2, or more precisely the *standard deviation* σ_X, describes the spread or dispersion of the random variable X around its mean μ_X. It is straightforward to show that

$$\sigma_X^2 = \mathbb{E}\left[(X - \mu_X)^2\right]$$
$$= \mathbb{E}[X^2 - 2\mu_X X + \mu_X^2] = \mathbb{E}[X^2] - 2\mu_X^2 + \mu_X^2$$
$$= \mathbb{E}[X^2] - \mu_X^2.$$

Often this is memorized as "variance equals the mean of the square minus the square of the mean."

Example 2.9 (Moments of a Gaussian random variable). If a random variable X has a normal distribution with density (2.5), then we have

$$\mu_X = \mathbb{E}[X] = \mu,$$
$$\sigma_X^2 = \text{var}(X) = \sigma,$$
$$\mathbb{E}\left[(X - \mu)^{2n-1}\right] = 0, \quad n = 1, 2, \ldots,$$
$$\mathbb{E}\left[(X - \mu)^{2n}\right] = 1 \cdot 3 \cdot 5 \cdots (2n - 1) \cdot \sigma^{2n}, \quad n = 1, 2, \ldots.$$

Therefore, the mean and variance of a normal distribution can completely characterize all of its moments.

2.2.4 Moment-Generating Function

Definition 2.10. *The moment-generating function for a random variable $X(\omega)$ is defined as $m_X(t) \triangleq \mathbb{E}[e^{tX}]$. It exists if there exists $b > 0$ such that $m_X(t)$ is finite for $|t| \leq b$. The reason $m_X(t)$ is called a moment-generating function is because*

$$\mu_k = \left.\frac{d^k m_X(t)}{dt^k}\right|_{t=0}, \quad k = 0, 1, \ldots,$$

where $\mu_k = \mathbb{E}[X^k]$ is the kth moment of X. The relationship can be seen as follows:

$$m_X(t) = \mathbb{E}\left[e^{tX}\right] = \int e^{tx} p_X(x) dx$$

$$= \int \sum_{k=0}^{\infty} \frac{tx^k}{k!} p_X(x) dx = \sum_{k=0}^{\infty} \frac{1}{k!} \int t^k x^k p_X(x) dx$$

$$= \sum_{k=0}^{\infty} \frac{t^k \mu_k}{k!} = \mu_0 + t\mu_1 + \frac{t^2}{2}\mu_2 + \cdots . \tag{2.6}$$

If $X \sim \mathcal{N}(0, \sigma^2)$ is a Gaussian random variable, then $m_X(t) = e^{\sigma^2 t^2/2}$.

2.2.5 Random Number Generation

One of the basis tasks in stochastic simulations is to generate a sequence of random numbers satisfying a desired probability distribution. To this end, one first seeks to generate a random sequence with a common uniform distribution in $(0, 1)$. There are many available algorithms that have been well studied. The algorithms in practical implementations are all deterministic (typically using recursion) and can therefore at best mimic properties of uniform random variables. For this reason, the sequence of outputs is called a sequence of *pseudorandom numbers*. Despite some defects in the early work, the algorithms for generating pseudorandom numbers have been much improved. The readers of this book will do well with existing software, which is fast and certainly faster than self-made high-level-language routines, and will seldom be able to improve it substantially. Therefore, we will not spend time on this subject, and we refer interested readers to references such as [38, 59, 65, 87].

For nonuniform random variables, the most straightforward technique is via the inversion of a distribution function. Let $F_X(x) = P(X \leq x)$ be the distribution function of X. For the simple case where F_X is strictly increasing and continuous, then $x = F_X^{-1}(u)$ is the unique solution of $F_X(x) = u, 0 < u < 1$. For distributions with nonconnected support or jumps, F_X is not strictly increasing, and more care is needed to find its inverse. We choose the left-continuous version

$$F_X^{-1}(u) \triangleq \inf\{x : F_X(x) \geq u\}. \tag{2.7}$$

We state here the following results that justify the inversion method for generating nonuniform random numbers.

Proposition 2.11. *Let $F_X(x) = P(X \leq x)$ be the distribution function of X. Then the following results hold.*

- $u \leq F_X(x) \iff F_X^{-1}(u) \leq x.$
- *If U is uniform in $(0, 1)$, then $F_X^{-1}(U)$ has distribution function F_X.*
- *If F_X is continuous, then $F_X(x)$ is uniform in $(0, 1)$.*

The part that is mainly used in simulation is the second statement, which allows us to generate X as $F_X^{-1}(U)$. The most common case is an F_X that is continuous and strictly increasing on an interval.

Example 2.12 (Exponential random variable). Let X be a random variable with *exponential distribution* whose probability density is $f_X(x) = ae^{-ax}$. Here $a > 0$ is a parameter that is often called the rate. It is easy to see that $\mathbb{E}[X] = 1/a$, the inverse of the rate. Then the inverse of F_X is $F_X^{-1}(u) = -\log(1 - x)/a$. So we can generate X by inversion: $X = -\log(1 - U)/a$. In practice, one often uses the equivalent $X = -\log(U)/a$.

A main limitation of inversion is that quite often F_X^{-1} is not available in explicit form, for example, when X is a normal random variable. Sometimes approximations are used.

Example 2.13 (Approximate inversion of normal distribution). Let X be a Gaussian random variable with zero mean and unit variance; i.e., $X \sim \mathcal{N}(0, 1)$. Its probability density function is $f_X(x) = \frac{1}{\sqrt{2\pi}}e^{-x^2/2}$, and there is no explicit formula for $F_X(x)$. The following two approximations to F_X^{-1} are quite simple and accurate.

$$F_X^{-1}(u) \approx \text{sign}(u - 1/2)\left(t - \frac{c_0 + c_1 t + c_2 t^2}{1 + d_1 t + d_2 t^2 + d_3 t^3}\right), \quad (2.8)$$

where $t = (-\ln[\min(u, (1 - u))]^2)^{1/2}$ and $c_0 = 2.515517$, $c_1 = 0.802853$, $d_1 = 1.432788$, $d_2 = 0.189269$, $d_3 = 0.001308$. The formula has absolute error less than 4.5×10^{-4} ([50]). Or,

$$F_X^{-1}(u) \approx y + \frac{p_0 + p_1 y + p_2 y^2 + p_3 y^3 + p_4 y^4}{q_0 + q_1 y + q_2 y^2 + q_3 y^3 + q_4 y^4}, \quad 0.5 < u < 1, \quad (2.9)$$

where $y = \sqrt{-2\log(1 - u)}$, the case of $0 < u < 0.5$ is handled by symmetry, and p_k, q_k are given in the accompanying table.

k	p_k	q_k
0	−0.322232431088	0.099348462606
1	−1	0.588581570495
2	−0.342242088547	0.531103462366
3	−0.0204231210245	0.10353775285
4	−0.0000453642210148	0.0038560700634

Other techniques exist for the generation of nonuniform random variables, most notably acceptance-rejection algorithms. We will not engage in more in-depth discussions here and refer interested readers to comprehensive references such as [25, 38, 52].

2.3 RANDOM VECTORS

We say $\mathbf{X} = (X_1, \ldots, X_n)$ is an n-dimensional *random vector* if its components $X_1, \ldots, X_n$ are one-dimensional real-valued random variables. Therefore, a random vector is nothing but a collection of a finite number of random variables. Similarly, we can also define concepts such as distribution function, moments, etc.

Definition 2.14. *The collection of the probabilities*

$$F_{\mathbf{X}}(\mathbf{x}) = P(X_1 \leq x_1, \ldots, X_n \leq x_n), \qquad \mathbf{x} = (x_1, \ldots, x_n) \in \mathbb{R}^n, \qquad (2.10)$$

is the distribution function $F_{\mathbf{X}}$ of $\mathbf{X}$.

If the distribution of a random vector $\mathbf{X}$ has a density $f_{\mathbf{X}}$, we can represent the distribution function $F_{\mathbf{X}}$ as

$$F_{\mathbf{X}}(x_1, \ldots, x_n) = \int_{-\infty}^{x_1} \cdots \int_{-\infty}^{x_n} f_{\mathbf{X}}(y_1, \ldots, y_n) dy_1 \cdots dy_n,$$

where the density is a function satisfying

$$f_{\mathbf{X}}(\mathbf{x}) \geq 0, \qquad \forall \mathbf{x} \in \mathbb{R}^n,$$

and

$$\int_{-\infty}^{\infty} \cdots \int_{-\infty}^{\infty} f_{\mathbf{X}}(y_1, \ldots, y_n) dy_1 \cdots dy_n = 1.$$

If a vector $\mathbf{X}$ has density $f_{\mathbf{X}}$, then all of its components X_i, the vectors of the pairs (X_i, X_j), triples (X_i, X_j, X_k), etc., have a density. They are called *marginal densities*. For example,

$$f_{X_i}(x_i) = \int_{-\infty}^{\infty} \cdots \int_{-\infty}^{\infty} f_{\mathbf{X}}(y_1, \ldots, y_n) dy_1 \cdots dy_{i-1} dy_{i+1} \cdots dy_n.$$

The important statistical quantities of a random vector include its expectation, variance, and covariance. The *expectation* or *mean value* of a random vector $\mathbf{X}$ is given by

$$\mu_{\mathbf{X}} = \mathbb{E}[\mathbf{X}] = (\mathbb{E}[X_1], \ldots, \mathbb{E}[X_n]).$$

The *covariance matrix* of $\mathbf{X}$ is defined as

$$\mathbf{C}_{\mathbf{X}} = (\mathrm{cov}(X_i, X_j))_{i,j=1}^n, \qquad (2.11)$$

where

$$\mathrm{cov}(X_i, X_j) = \mathbb{E}[(X_i - \mu_{X_i})(X_j - \mu_{X_j})] = \mathbb{E}(X_i X_j) - \mu_{X_i} \mu_{X_j}, \qquad (2.12)$$

is the *covariance* of X_i and X_j. Note that $\mathrm{cov}(X_i, X_i) = \sigma_{X_i}^2$.

It is also convenient to standardize covariances by dividing the random variables by their standard deviations. The resulting quantity

$$\mathrm{corr}(X_1, X_2) = \frac{\mathrm{cov}(X_1, X_2)}{\sigma_{X_1} \sigma_{X_2}} \qquad (2.13)$$

is the *correlation coefficient*. An immediate fact following the Cauchy-Schwarz inequality is

$$-1 \leq \mathrm{corr}(X_1, X_2) \leq 1. \qquad (2.14)$$

We say the two random variables are *uncorrelated* if $\mathrm{corr}(X_1, X_2) = 0$, and *strongly correlated* if $|\mathrm{corr}(X_1, X_2)| \approx 1$.

Example 2.15 (Gaussian random vector). A *Gaussian* or *normal random vector* has a Gaussian or normal distribution. The n-dimensional Gaussian distribution is defined by its density

$$f_{\mathbf{X}}(\mathbf{x}) = \frac{1}{(2\pi)^{n/2}(\det \mathbf{C_X})^{1/2}} \exp\left\{ -\frac{1}{2}(\mathbf{x} - \mu_{\mathbf{X}})\mathbf{C_X}^{-1}(\mathbf{x} - \mu_{\mathbf{X}})^T \right\}, \qquad (2.15)$$

where $\mu_{\mathbf{X}} \in \mathbb{R}^n$ is the expectation of $\mathbf{X}$ and $\mathbf{C_X}$ is the covariance matrix. Thus, the density of a Gaussian vector (hence its distribution) is completely determined via its expectation and covariance matrix. If $\mathbf{C_X} = \mathbf{I}_n$, the n-dimensional identity matrix, the components are called *uncorrelated* and the density becomes the product of n normal densities:

$$f_{\mathbf{X}}(\mathbf{x}) = f_{X_1}(x_1) \cdots f_{X_n}(x_n),$$

where $f_{X_i}(x_i)$ is the normal density of $\mathcal{N}(\mu_{X_i}, \sigma_{X_i}^2)$. An important and appealing property of Gaussian random vectors is that they remain Gaussian under linear transformation.

Theorem 2.16. *Let* $\mathbf{X} = (X_1, \ldots, X_n)$ *be a Gaussian random vector with distribution* $\mathcal{N}(\mu, \mathbf{C})$ *and let* $\mathbf{A}$ *be an* $m \times n$ *matrix. Then* $\mathbf{A}\mathbf{X}^T$ *has an* $\mathcal{N}(\mathbf{A}\mu^T, \mathbf{A}\mathbf{C}\mathbf{A}^T)$ *distribution.*

The proof is left as an exercise.

2.4 DEPENDENCE AND CONDITIONAL EXPECTATION

Intuitively, two random events are called *independent* if the outcome of one event does not influence the outcome of the other. More precisely, we state the following.

Definition 2.17. *Two events* A_1 *and* A_2 *are independent if*

$$P(A_1 \cap A_2) = P(A_1)P(A_2).$$

Definition 2.18. *Two random variables* X_1 *and* X_2 *are independent if*

$$P(X_1 \in B_1, X_2 \in B_2) = P(X_1 \in B_1)P(X_2 \in B_2)$$

for all suitable subsets B_1 *and* B_2 *of* $\mathbb{R}$. *This means that the events* $\{X_1 \in B_1\}$ *and* $\{X_2 \in B_2\}$ *are independent.*

Alternatively, one can define independence via distribution functions and densities. The random variables $X_1, \ldots, X_n$ are independent if and only if their joint distribution function can be written as

$$F_{X_1, \ldots, X_n}(x_1, \ldots, x_n) = F_{X_1}(x_1) \cdots F_{X_n}(x_n), \qquad (x_1, \ldots, x_n) \in \mathbb{R}^n.$$

If the random vector $\mathbf{X} = (X_1, \ldots, X_n)$ has density $f_{\mathbf{X}}$, then $X_1, \ldots, X_n$ are independent if and only if

$$f_{X_1, \ldots, X_n}(x_1, \ldots, x_n) = f_{X_1}(x_1) \cdots f_{X_n}(x_n), \qquad (x_1, \ldots, x_n) \in \mathbb{R}^n. \qquad (2.16)$$

It follows immediately that if $X_1, \ldots, X_n$ are independent, then for any real-valued functions $g_1, \ldots, g_n$,

$$\mathbb{E}[g_1(X_1) \cdots g_n(X_n)] = \mathbb{E}[g_1(X_1)] \cdots \mathbb{E}[g_n(X_n)],$$

provided the considered expectations are all well defined. Hence, if X_1 and X_2 are independent, then

$$\text{corr}(X_1, X_2) = \text{cov}(X_1, X_2) = 0.$$

This implies that *independent random variables are uncorrelated*. The converse is in general not true.

Example 2.19. Let X be a standard normal random variable. Since X is symmetric (its density is an even function), so is X^3. Therefore, both X and X^3 have expectation zero. Thus,

$$\text{cov}(X, X^2) = \mathbb{E}[X^3] - \mathbb{E}[X]\mathbb{E}[X^2] = 0,$$

which implies that X and X^2 are uncorrelated. However, they are clearly dependent, in the sense that knowing X determines X^2 completely. For example, since $\{X \in [-1, 1]\} = \{X^2 \in [0, 1]\}$, we have

$$
\begin{aligned}
P(X \in [-1, 1], X^2 \in [0, 1]) &= P(X \in [-1, 1]) \\
&> P(X \in [-1, 1]) P(X^2 \in [0, 1]) \\
&= (P(X \in [-1, 1]))^2.
\end{aligned}
$$

Example 2.20. Let $\mathbf{X}$ be an n-dimensional Gaussian random vector with density (2.15). The components are uncorrelated when $\text{corr}(X_i, X_j) = \text{cov}(X_i, X_j) = 0$ for $i \neq j$. This means that the correlation matrix is diagonal. In this case, the density of $\mathbf{X}$ can be written in the product form of (2.16), and therefore the components are independent. Thus, *uncorrelation and independence are equivalent notions for Gaussian distributions.*

Another concept describing relations among random events or random variables is the *conditional expectation*. This is an extremely important subject in probability theory, though it is not of as much significance in this book.

From elementary probability theory, the *conditional probability of A given B* is

$$P(A|B) = \frac{P(A \cap B)}{P(B)}.$$

Clearly,

$$P(A|B) = P(A) \quad \text{if and only if } A \text{ and } B \text{ are independent.}$$

Given that $P(B) > 0$, we can define the *conditional distribution function of a random variable X given B*,

$$F_X(x|B) = \frac{P(X \leq x, B)}{P(B)}, \quad x \in \mathbb{R},$$

and also the *conditional expectation of X given B*,

$$\mathbb{E}[X|B] = \frac{\mathbb{E}[XI_B]}{P(B)}, \tag{2.17}$$

where

$$I_B(\omega) = \begin{cases} 1, & \omega \in B, \\ 0, & \omega \notin B, \end{cases}$$

denotes the *indicator function of the event B*. For the moment let us assume that $\Omega = \mathbb{R}$. If X is a discrete random variable, then (2.17) becomes

$$\mathbb{E}[X|B] = \sum_{k=1}^{\infty} x_k \frac{P(\{\omega : X(\omega) = x_k\} \cap B)}{P(B)} = \sum_{k=1}^{\infty} x_k P(X = x_k|B).$$

If X has density f_X, then (2.17) becomes

$$\mathbb{E}[X|B] = \frac{1}{P(B)} \int x I_B(x) f_X(x) dx = \frac{1}{P(B)} \int_B x f_X(x) dx.$$

2.5 STOCHASTIC PROCESSES

In many physical systems, randomness varies either continuously or discretely over physical space and/or time. Therefore, it is necessary to study the distribution and evolution of the random variables that describe the randomness as functions of space and/or time. A mathematical model for describing this is called a stochastic process or a random process.

A *stochastic process* is a collection of random variables

$$(X_t, t \in T) = (X_t(\omega), t \in T, \omega \in \Omega)$$

defined on some space Ω. Here t is the index of the random variable X. The index set T can be an interval, e.g., $T = [a, b], [a, b)$ or $[a, \infty)$ for $a < b$. Then X is a *continuous* process. If T is a finite or countably infinite set, then X is a discrete process. Very often the index t of X_t is referred to as *time*. However, one should keep in mind that it is merely an index and can be a space location as well.

A stochastic process X can be considered as a function of two variables.

- For a fixed index t, it is a random variable:

$$X_t = X_t(\omega), \qquad \omega \in \Omega.$$

- For a fixed random outcome $\omega \in \Omega$, it is a function of the index (time):

$$X_t = X_t(\omega), \qquad t \in T.$$

This is called a *realization*, a *trajectory*, or a *sample path* of the process X_t.

It is then natural to seek the statistics of the stochastic process. The task is more complicated than that for a random vector. For example, a process X_t with an infinite index set T is an infinite-dimensional object. Mathematical care is needed for such objects. A simpler approach, which suits practical needs well, is to interpret the process as a collection of random vectors. In this way, we study the *finite-dimensional distributions* of the stochastic process X_t, which are defined as the distributions of the finite-dimensional vectors

$$\left(X_{t_1}, \dots, X_{t_n} \right), \qquad t_1, \dots, t_n \in T,$$

for all possible choices of the index $t_1, \ldots, t_n \in T$ and every $n \geq 1$. These are easier to study and indeed determine the distribution of X_t. In this sense, we refer to a collection of finite-dimensional distributions as the distribution of the stochastic process.

A stochastic process $X = (X_t, t \in T)$ can be considered a collection of random vectors $(X_{t_1}, \ldots, X_{t_n})$ for $t_1, \ldots, t_n \in T$ and $n \geq 1$. We can then extend the definitions for expectation and covariance matrices for the random vector to the process and consider these quantities as functions of $t \in T$. The *expectation function* of X is given by

$$\mu_X(t) = \mu_{X_t} = \mathbb{E}[X_t], \qquad t \in T.$$

The *covariance function* of X is given by

$$C_X(t, s) = \mathrm{cov}(X_t, X_s) = \mathbb{E}[(X_t - \mu_X(t))(X_s - \mu_X(s))], \qquad t, s \in T.$$

The *variance* of X is given by

$$\sigma_X^2(t) = C_X(t, t) = \mathrm{var}(X_t), \qquad t \in T.$$

These are obviously deterministic quantities. The expectation function $\mu_X(t)$ is a deterministic path around which the sample paths of X are concentrated. *Note that in many cases $\mu_X(t)$ may not be a realizable sample path.* The variance function can be considered a measure of the spread of the sample paths around the expectation. The covariance function is a measure of dependence on the process X.

The process $X = (X_t, t \in T)$ is called *strictly stationary* if the finite-dimensional distributions of the process are invariant under shifts of the index t:

$$(X_{t_1}, \ldots, X_{t_n}) \overset{d}{=} (X_{t_1+h}, \ldots, X_{t_n+h})$$

for all possible choices of indices $t_1, \ldots, t_n \in T$, $n \geq 1$ and h such that $t_1 + h, \ldots, t_n + h \in T$. Here $\overset{d}{=}$ stands for the identity of the distributions; see the following section for the definition. In practice, a weaker version of stationarity is often adopted. A process X is called *stationary in the wide sense* or *second-order stationary* if its expectation is a constant and covariance function $C_X(t, s)$ depends only on the distance $|t - s|$. For a Gaussian process, since its mean and covariance function can fully characterize the distribution of the process, the two concepts for stationarity become equivalent.

A large class of (extremely) useful processes can be constructed by imposing a stationary (strictly or in the wide sense) condition on the increment of the processes. Examples include the homogeneous Poisson process and Brownian motion. Extensive mathematical analysis has been conducted on such processes by using nonelementary facts from measure theory and functional analysis. This, however, is not the focus of this book. We refer interested readers to the many excellent books such as [36, 55].

2.6 MODES OF CONVERGENCE

We now introduce concepts of main modes of convergence for a sequence of random variables $X_1, X_2, \ldots$.

Definition 2.21 (Convergence in distribution). *The sequence* $\{X_n\}$ *converges in distribution or converges weakly to the random variable X, written as* $X_n \xrightarrow{d} X$, *if for all bounded and continuous functions* f,

$$\mathbb{E}[f(X_n)] \to \mathbb{E}[f(X)], \qquad n \to \infty.$$

Note that $X_n \xrightarrow{d} X$ holds if and only if for all continuous points x of the distribution function F_X the relation

$$F_{X_n}(x) \to F_X(x), \qquad n \to \infty,$$

is satisfied. If F_X is continuous, this can be strengthened to uniform convergence:

$$\sup_x |F_{X_n}(x) - F_X(x)| \to 0, \qquad n \to \infty.$$

We state the following useful theorem without proving it.

Theorem 2.22. *Let* X_n *and* X *be random variables with moment-generating functions* $m_{X_n}(t)$ *and* $m_X(t)$, *respectively. If*

$$\lim_{n \to \infty} m_{X_n}(t) = m_X(t), \qquad \forall t,$$

then $X_n \xrightarrow{d} X$ *as* $n \to \infty$.

Definition 2.23 (Convergence in probability). *The sequence* $\{X_n\}$ *converges in probability to X, written as* $X_n \xrightarrow{P} X$, *if for all positive* ϵ,

$$P(|X_n - X|) > \epsilon) \to 0, \qquad n \to \infty.$$

Convergence in probability implies convergence in distribution. The converse is true if and only if $X = x$ for some constant x.

Definition 2.24 (Almost sure convergence). *The sequence* $\{X_n\}$ *converges almost surely (a.s.), or with probability 1, to the random variable X, written as* $X_n \xrightarrow{a.s.} X$, *if the set of* ω *with*

$$X_n(\omega) \to X(\omega), \qquad n \to \infty,$$

has probability 1.

This implies that

$$P(X_n \to X) = P(\{\omega : X_n(\omega) \to X(\omega)\}) = 1.$$

Convergence with probability 1 implies convergence in probability, hence convergence in distribution. Convergence in probability does not imply a.s. convergence. However, $X_n \xrightarrow{P} X$ implies that $X_{n_k} \xrightarrow{a.s.} X$ for a suitable subsequence $\{X_{n_k}\}$.

Definition 2.25 (L^p convergence). *Let $p > 0$. The sequence $\{X_n\}$ converges in L^p, or in the pth mean, to X, written as $X_n \overset{L^p}{\to} X$, if $\mathbb{E}[|X_n|^p + |X|^p] < \infty$ for all n and*

$$\mathbb{E}[|X_n - X|^p] \to 0, \qquad n \to \infty.$$

The well-known Markov's inequality ensures that $P(|X_n - X| > \epsilon) \leq \epsilon^{-p}\mathbb{E}[|X_n - X|^p]$ for positive p and ϵ. Thus, $X_n \overset{L^p}{\to} X$ implies $X_n \overset{P}{\to} X$. The converse is in general not true.

For $p = 2$, we say that X_n converges to X in mean square. Mean-square convergence is convergence in the Hilbert space

$$L^2 = L^2(\Omega, \mathcal{F}, P) = \{X : \mathbb{E}[X^2] < \infty\}$$

endowed with the inner product $\langle X, Y \rangle = \mathbb{E}[XY]$ and the norm $\|X\| = \sqrt{\langle X, X \rangle}$.

Convergence in distribution and convergence in probability are often referred to as *weak convergence*, whereas a.s. convergence and L^p convergence are often referred to as *strong convergence*.

2.7 CENTRAL LIMIT THEOREM

The celebrated central limit theorem (CLT) plays a central role in many aspects of probability analysis. Here we state a version of the CLT that suits the exposition of this book.

Theorem 2.26. *Let $X_1, X_2, \ldots, X_n$ be independent and identically distributed (i.i.d.) random variables with $\mathbb{E}[X_i] = \mu$ and $var(X_i) = \sigma^2 < \infty$. Let*

$$\overline{X} = \frac{1}{n} \sum_{i=1}^{n} X_i$$

and let

$$U_n = \sqrt{n} \left(\frac{\overline{X} - \mu}{\sigma} \right).$$

Then the distribution function of U_n converges to an $\mathcal{N}(0, 1)$ distribution function as $n \to \infty$.

Proof. For all i, let $Z_i = \frac{X_i - \mu}{\sigma}$; then $\mathbb{E}[Z_i] = 0$ and $var(Z_i) = 1$. Also, let $\mu_3 = \mathbb{E}[Z_i^3]$ for all i. Then, by following the property of moment-generating function (2.6), we have

$$m_{Z_i}(t) = 1 + \frac{t^2}{2} + \frac{t^3}{3!}\mu_3 + \cdots.$$

By definition,

$$U_n = \sqrt{n} \left(\frac{\overline{X} - \mu}{\sigma} \right) = \frac{1}{\sqrt{n}} \left(\frac{\sum_i X_i - n\mu}{\sigma} \right) = \frac{1}{\sqrt{n}} \sum_i Z_i.$$

We immediately have

$$m_{U_n}(t) = \prod_i m_{Z_i}\left(\frac{t}{\sqrt{n}}\right) = \left(m_{Z_i}\left(\frac{t}{\sqrt{n}}\right)\right)^n = \left(1 + \frac{t^2}{2n} + \frac{t^3}{3!n^{3/2}}\mu_3 + \cdots\right)^n$$

and

$$\ln\left(m_{U_n}(t)\right) = n\ln\left(1 + \frac{t^2}{2n} + \frac{t^3}{3!n^{3/2}}\mu_3 + \cdots\right).$$

Letting $z = \frac{t^2}{2n} + \frac{t^3}{3!n^{3/2}}\mu_3 + \cdots$ and using $\ln(1+z) = z - \frac{z^2}{2} + \frac{z^3}{3} - \frac{z^4}{4} + \cdots$, we have

$$\ln(m_{U_n}(t)) = n\left(z - \frac{z^2}{2} + \cdots\right) = n\left(\frac{t^2}{2n} + \frac{t^3\mu_3}{3!n^{3/2}} + \cdots\right).$$

It is obvious that

$$\lim_{n\to\infty} m_{U_n}(t) = \frac{t^2}{2},$$

which is the moment-generating function of a unit Gaussian random variable $\mathcal{N}(0, 1)$. The theorem is then proved by virtue of theorem 2.22.

This immediately implies that the numerical average of a set of i.i.d. random variables $\{X_i\}_{i=1}^n$ will converge, as n is increased, to a Gaussian distribution $\mathcal{N}(\mu, \sigma^2/n)$, where μ and σ^2 are the mean and variance of the i.i.d random variables.

Chapter Three

Survey of Orthogonal Polynomials and Approximation Theory

In this chapter we review the basic aspects of orthogonal polynomials and approximation theory. We will focus exclusively on the univariate case, that is, polynomials and approximations on the real line, to establish the fundamentals of the theories.

3.1 ORTHOGONAL POLYNOMIALS

We first review the basics of orthogonal polynomials, which play a central role in modern approximation theory. The material is kept to a minimum to satisfy the needs of this book. More in-depth discussions of the properties of orthogonal polynomials can be found in many standard books such as [10, 19, 100].

From here on we adopt the standard notation by letting $\mathbb{N}$ be the set of positive integers and letting $\mathbb{N}_0$ be the set of nonnegative integers. We also let $\mathcal{N} = \mathbb{N}_0 = \{0, 1, \dots\}$ or $\mathcal{N} = \{0, 1, \dots, N\}$ be an index set for a finite nonnegative integer N.

3.1.1 Orthogonality Relations

A general polynomial of degree n takes the form

$$Q_n(x) = a_n x^n + a_{n-1} x^{n-1} + \cdots + a_1 x + a_0, \qquad a_n \neq 0, \qquad (3.1)$$

where a_n is the *leading coefficient* of the polynomial. We denote by

$$P_n(x) = \frac{Q_n(x)}{a_n} = x^n + \frac{a_{n-1}}{a_n} x^{n-1} + \cdots + \frac{a_1}{a_n} x + \frac{a_0}{a_n}$$

the *monic version* of this polynomial, i.e., the one with the leading coefficients equal to 1.

A system of polynomials $\{Q_n(x), n \in \mathcal{N}\}$ is an *orthogonal system of polynomials* with respect to some real positive measure α if the following orthogonality relations hold:

$$\int_S Q_n(x) Q_m(x) d\alpha(x) = \gamma_n \delta_{mn}, \qquad m, n \in \mathcal{N}, \qquad (3.2)$$

where $\delta_{mn} = 0$ if $m \neq n$ and $\delta_{mn} = 1$ if $m = n$, is the Kronecker delta function, S is the support of the measure α, and γ_n are positive constants often termed *normalization constants*. Obviously,

$$\gamma_n = \int_S Q_n^2(x) d\alpha(x), \qquad n \in \mathcal{N}.$$

If $\gamma_n = 1$, the system is *orthonormal*. Note that by defining $\widetilde{Q}_n(x) = Q_n(x)/\sqrt{\gamma_n}$, the system $\{\widetilde{Q}_n\}$ is orthonormal.

The measure α usually has a density $w(x)$ or is a discrete measure with weight w_i at the points x_i. The relations (3.2) then become

$$\int_S Q_n(x)Q_m(x)w(x)dx = \gamma_n\delta_{mn}, \qquad m, n \in \mathcal{N}, \tag{3.3}$$

in the former case and

$$\sum_i Q_n(x_i)Q_m(x_i)w_i = \gamma_n\delta_{mn}, \qquad m, n \in \mathcal{N}, \tag{3.4}$$

in the latter case where it is possible that the summation is an infinite one.

If we define a weighted inner product

$$(u, v)_{d\alpha} = \int_S u(x)v(x)d\alpha(x), \tag{3.5}$$

which in continuous cases takes the form

$$(u, v)_w = \int_S u(x)v(x)w(x)dx \tag{3.6}$$

and in discrete cases takes the form

$$(u, v)_w = \sum_i u(x_i)v(x_i)w_i, \tag{3.7}$$

then the orthogonality relations can be written as

$$(Q_m Q_n)_w = \gamma_n\delta_{mn} \qquad m, n \in \mathcal{N}, \tag{3.8}$$

where

$$\gamma_n = (Q_n, Q_n)_w = \|Q_n\|_w^2, \qquad n \in \mathcal{N}. \tag{3.9}$$

3.1.2 Three-Term Recurrence Relation

It is well known that all orthogonal polynomials $\{Q_n(x)\}$ on the real line satisfy a three-term recurrence relation

$$-x Q_n(x) = b_n Q_{n+1}(x) + a_n Q_n(x) + c_n Q_{n-1}(x), \qquad n \geq 1, \tag{3.10}$$

where $b_n, c_n \neq 0$ and $c_n/b_{n-1} > 0$. Along with $Q_{-1}(x) = 0$ and $Q_0(x) = 1$, the three-term recurrence defines the polynomial system completely. Often the relation is written in a different form,

$$Q_{n+1}(x) = (A_n x + B_n)Q_n(x) - C_n Q_{n-1}(x), \qquad n \geq 0,$$

and Favard proved the following converse result ([19]).

Theorem 3.1 (Favard's Theorem). *Let A_n, B_n, and C_n be arbitrary sequences of real numbers and let $\{Q_n(x)\}$ be defined by the recurrence relation*

$$Q_{n+1}(x) = (A_n x + B_n)Q_n(x) - C_n Q_{n-1}(x), \qquad n \geq 0,$$

together with $Q_0(x) = 1$ and $Q_{-1}(x) = 0$. Then the $\{Q_n(x)\}$ are a system of orthogonal polynomials if and only if $A_n \neq 0$, $C_n \neq 0$, and $C_n A_n A_{n-1} > 0$ for all n.

3.1.3 Hypergeometric Series and the Askey Scheme

Most orthogonal polynomials can be expressed in a unified way by using *hypergeometric series* and incorporated in the *Askey scheme*. To this end, we first define the *Pochhammer symbol* $(a)_n$ as

$$(a)_n = \begin{cases} a, & n = 0, \\ a(a+1)\cdots(a+n-1), & n = 1, 2, \ldots. \end{cases} \tag{3.11}$$

If $a \in \mathbb{N}$ is an integer, then

$$(a)_n = \frac{(a+n-1)!}{(a-1)!}, \qquad n > 0,$$

and for general $a \in \mathbb{R}$,

$$(a)_n = \frac{\Gamma(a+n)}{\Gamma(a)}, \qquad n > 0.$$

The *generalized hypergeometric series* $_rF_s$ is defined by

$$_rF_s(a_1, \ldots, a_r; b_1, \ldots, b_s; z) = \sum_{k=0}^{\infty} \frac{(a_1)_k \cdots (a_r)_k}{(b_1)_k \cdots (b_s)_k} \frac{z^k}{k!}, \tag{3.12}$$

where $b_i \neq 0, -1, -2, \ldots$, for all $i = 1, \ldots, s$. There are r parameters in the numerator and s parameters in the denominator. Clearly, the orders of the parameters are immaterial.

For example,

$$_0F_0(\ ;\ ; z) = \sum_{k=0}^{\infty} \frac{z^k}{k!}$$

is the power series for an exponential function.

When the series is infinite, its radius of convergence ρ is

$$\rho = \begin{cases} \infty, & r < s + 1, \\ 1, & r = s + 1, \\ 0, & r > s + 1. \end{cases}$$

If one of the numerator parameters a_i, $i = 1, \ldots, r$, is a negative integer, say, $a_1 = -n$, the series terminates because $(a_1)_k = (-n)_k = 0$, for $k = n + 1, n + 2, \ldots$, and becomes

$$_rF_s(a_1, \ldots, a_r; b_1, \ldots, b_s; z) = \sum_{k=0}^{n} \frac{(-n)_k \cdots (a_r)_k}{(b_1)_k \cdots (b_s)_k} \frac{z^k}{k!}. \tag{3.13}$$

This is a polynomial of degree n.

The Askey scheme, which can be represented as a tree structure as shown in figure 3.1, classifies the hypergeometric orthogonal polynomials and indicates the limit relations between them. The tree starts with Wilson polynomials and Racah polynomials at the top. They both belong to class $_4F_3$ of hypergeometric orthogonal polynomials. Wilson polynomials are continuous polynomials and Racah polynomials are discrete. The lines connecting different polynomials denote the limit transition relationships between them, which implies that polynomials at the lower ends

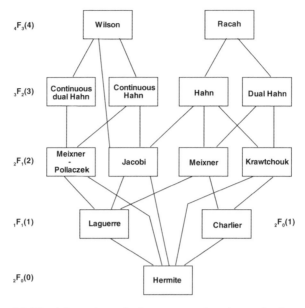

Figure 3.1 The Askey scheme for hypergeometric orthogonal polynomials.

of the lines can be obtained by taking the limit of some parameters in the polynomials at the upper ends. For example, the limit relation between Jacobi polynomials $P_n^{(\alpha,\beta)}(x)$ and Hermite polynomials $H_n(x)$ is

$$\lim_{\alpha \to \infty} \alpha^{-\frac{1}{2}n} P_n^{(\alpha,\alpha)} \left(\frac{x}{\sqrt{\alpha}} \right) = \frac{H_n(x)}{2^n n!},$$

and that between Meixner polynomials $M_n(x; \beta, c)$ and Charlier polynomials $C_n(x; a)$ is

$$\lim_{\beta \to \infty} M_n \left(x; \beta, \frac{a}{a + \beta} \right) = C_n(x; a).$$

For a detailed account of hypergeometric polynomials and the Askey scheme, the interested reader should consult [60] and [91].

3.1.4 Examples of Orthogonal Polynomials

Here we present several orthogonal polynomials that will be used extensively in this book. The focus is on continuous polynomials. More specifically, we discuss Legendre polynomials defined on $[-1, 1]$, Hermite polynomials defined on $\mathbb{R}$, and Laguerre polynomials defined on $[0, \infty)$. These correspond to polynomials with support on a bounded interval (with proper scaling), the entire real line, and the half real line, respectively.

3.1.4.1 Legendre Polynomials

Legendre polynomials

$$P_n(x) = {}_2F_1\left(-n, n+1; 1; \frac{1-x}{2}\right) \tag{3.14}$$

satisfy

$$P_{n+1} = \frac{2n+1}{n+1}x\,P_n(x) - \frac{n}{n+1}P_{n-1}(x), \qquad n > 0, \tag{3.15}$$

and

$$\int_{-1}^{1} P_n(x)P_m(x)dx = \frac{2}{2n+1}\delta_{mn}. \tag{3.16}$$

Obviously the weight function in the orthogonality relation is a constant; i.e., $w(x) = 1$. The first few Legendre polynomials are

$$P_0(x) = 1, \quad P_1(x) = x, \quad P_2(x) = \frac{3}{2}x^2 - \frac{1}{2}, \quad \dots .$$

Legendre polynomials are a special case of Jacobi polynomials $P_n^{(\alpha,\beta)}(x)$ with parameters $\alpha = \beta = 0$. The details for Jacobi polynomials can be found in appendix A.

3.1.4.2 Hermite Polynomials

Hermite polynomials

$$H_n(x) = \left(\sqrt{2}x\right)^n {}_2F_0\left(-\frac{n}{n}, -\frac{n-1}{2}; \; ; -\frac{2}{x^2}\right) \tag{3.17}$$

satisfy

$$H_{n+1}(x) = x\,H_n(x) - n\,H_{n-1}(x), \qquad n > 0, \tag{3.18}$$

and

$$\int_{-\infty}^{\infty} H_m(x)H_n(x)w(x)dx = n!\delta_{mn}, \tag{3.19}$$

where

$$w(x) = \frac{1}{\sqrt{2\pi}}e^{-x^2/2}.$$

The first few Hermite polynomials are

$$H_0(x) = 1, \quad H_1(x) = x, \quad H_2(x) = x^2 - 1, \quad H_3(x) = x^3 - 3x, \quad \dots .$$

Note that the definition of $H_n(x)$ here is slightly different from the classical one used in the literature. Classical Hermite polynomials $\widetilde{H}_n(x)$ are often defined by

$$\widetilde{H}_{n+1}(x) = 2x\,\widetilde{H}_n(x) - 2n\,\widetilde{H}_{n-1}(x), \qquad n > 0,$$

and

$$\int_{-\infty}^{\infty} \tilde{H}_n(x) \tilde{H}_m(x) \tilde{w}(x) dx = 2^n n! \delta_{m,n},$$

where $\tilde{w}(x) = \frac{1}{\sqrt{\pi}} e^{-x^2/2}$. The two expressions are off by a scaling factor. We employ $H_n(x)$ here to facilitate the discussions associated with probability theory.

3.1.4.3 Laguerre Polynomials

Laguerre polynomials

$$L_n^{(\alpha)}(x) = \frac{(\alpha+1)_n}{n!} \, {}_1F_1(-n; \alpha+1; x), \qquad \alpha > -1, \qquad (3.20)$$

satisfy

$$(n+1)L_{n+1}^{(\alpha)}(x) = (-x+2n+\alpha+1)L_n^{(\alpha)}(x) - (n+\alpha)L_{n-1}^{(\alpha)}(x), \quad n > 0, \qquad (3.21)$$

and

$$\int_0^{\infty} L_m^{(\alpha)}(x) L_n^{(\alpha)}(x) w(x) dx = \frac{\Gamma(n+\alpha+1)}{n!} \delta_{m,n}, \qquad (3.22)$$

where

$$w(x) = e^{-x} x^{\alpha}.$$

Note that the sign of the leading coefficients of Laguerre polynomials flips with increasing degree.

3.2 FUNDAMENTAL RESULTS OF POLYNOMIAL APPROXIMATION

Let $\mathbb{P}_n$ be the linear space of polynomials of degree at most n; i.e.,

$$\mathbb{P}_n = \text{span}\{x^k : k = 0, 1, \ldots, n\}. \qquad (3.23)$$

We begin with a classical theorem by Weierstrass in approximation theory.

Theorem 3.2 (Weierstrass). *Let I be a bounded interval and let $f \in C^0(\bar{I})$. Then, for any $\epsilon > 0$, we can find $n \in \mathbb{N}$ and $p \in \mathbb{P}_n$ such that*

$$|f(x) - p(x)| < \epsilon, \qquad \forall x \in \bar{I}.$$

We skip the proof here. Interested readers can find the proof in various books on approximation theory, for example, [18, 105, 106]. This celebrated theorem states that any continuous function in a bounded closed interval can be uniformly approximated by polynomials. From this theorem, a large variety of sophisticated results have emerged. A natural question to ask is whether, among all the polynomials of degree less than or equal to a fixed integer n, it is possible to find one that best approximates a given continuous function f uniformly in $\bar{I}$. In other words, we would like to study the existence of $\phi_n(f) \in \mathbb{P}_n$ such that

$$\|f - \phi_n(f)\|_{\infty} = \inf_{\psi \in \mathbb{P}_n} \|f - \psi\|_{\infty}. \qquad (3.24)$$

This problem admits a unique solution, though the proof is very involved. An extensive and general treatise on this subject can be found in [105]. The nth-degree polynomial $\phi_n(f)$ is called the polynomial of *best uniform approximation* of f in $\bar{I}$. Following theorem 3.2, one immediately obtains

$$\lim_{n \to \infty} \|f - \phi_n(f)\|_\infty = 0.$$

3.3 POLYNOMIAL PROJECTION

From here on we do not restrict ourselves to bounded intervals and consider the general cases of $\bar{I}$, where $\bar{I} = [-1, 1]$, $\bar{I} = [0, \infty[$, or $\bar{I} = \mathbb{R}$.

Another best approximation problem can be formulated in terms of norms other than the infinity norm used in (3.24). To this end, we define, for a positive weight function $w(x)$, $x \in I$, the *weighted L^2 space* by

$$L_w^2(I) \triangleq \left\{ v : I \to \mathbb{R} \,\middle|\, \int_I v^2(x)w(x)dx < \infty \right\} \tag{3.25}$$

with the inner product

$$(u, v)_{L_w^2(I)} = \int_I u(x)v(x)w(x)dx, \qquad \forall u, v \in L_w^2(I), \tag{3.26}$$

and the norm

$$\|u\|_{L_w^2(I)} = \left(\int_I u^2(x)w(x)dx \right)^{1/2}. \tag{3.27}$$

Throughout this book, we will often use the simplified notation $(u, v)_w$ and $\|u\|_w$ to stand for $(u, v)_{L_w^2(I)}$ and $\|u\|_{L_w^2(I)}$, respectively, unless confusion would arise.

3.3.1 Orthogonal Projection

Let N be a fixed nonnegative integer and let $\{\phi_k(x)\}_{k=0}^N \subset \mathbb{P}_N$ be orthogonal polynomials of degree at most N with respect to the positive weight $w(x)$; i.e.,

$$(\phi_m(x), \phi_n(x))_{L_w^2(I)} = \|\phi_m\|_{L_w^2(I)}^2 \delta_{m,n}, \qquad 0 \le m, n \le N. \tag{3.28}$$

We introduce the *projection operator* $P_N : L_w^2(I) \to \mathbb{P}_N$ such that, for any function $f \in L_w^2(I)$,

$$P_N f \triangleq \sum_{k=0}^N \hat{f}_k \phi_k(x), \tag{3.29}$$

where

$$\hat{f}_k \triangleq \frac{1}{\|\phi_k\|_{L_w^2}^2}(f, \phi_k)_{L_w^2}, \qquad 0 \le k \le N. \tag{3.30}$$

Obviously, $P_N f \in \mathbb{P}_N$. It is called the *orthogonal projection* of f onto $\mathbb{P}_N$ via the inner product $(\cdot, \cdot)_{L_w^2}$, and $\{\hat{f}_k\}$ are the *(generalized) Fourier coefficients*.

The following trivial facts hold:

$$P_N f = f, \qquad \forall f \in \mathbb{P}_N,$$

$$P_N \phi_k = 0, \qquad \forall k > N.$$

Moreover, we have the following theorem.

Theorem 3.3. *For any $f \in L_w^2(I)$ and any $N \in \mathbb{N}_0$, $P_N f$ is the best approximation in the weighted L^2 norm (3.27) in the sense that*

$$\|f - P_N f\|_{L_w^2} = \inf_{\psi \in \mathbb{P}_N} \|f - \psi\|_{L_w^2}. \tag{3.31}$$

Proof. Any polynomial $\psi \in \mathbb{P}_N$ can be written in the form $\psi = \sum_{k=0}^{N} c_k \phi_k$ for some real coefficients c_k, $0 \le k \le N$. Minimizing $\|f - \psi\|_{L_w^2}$ is equivalent to minimizing $\|f - \psi\|_{L_w^2}^2$, whose derivatives are

$$\frac{\partial}{\partial c_j}\|f - \psi\|_{L_w^2}^2 = \frac{\partial}{\partial c_j}\left(\|f\|_{L_w^2}^2 - 2\sum_{k=0}^{N} c_k (f, \phi_k)_{L_w^2} + \sum_{k=0}^{N} c_k^2 \|\phi_k\|_{L_w^2}^2 \right)$$

$$= -2(f, \phi_j)_{L_w^2} + 2c_j \|\phi_j\|_{L_w^2}^2, \qquad 0 \le j \le N.$$

By setting the derivatives to zero, the unique minimum is attained when $c_j = \hat{f}_j$, $0 \le j \le N$, where $\hat{f}_j$ are the Fourier coefficients of f in (3.30). This completes the proof. $\qquad\blacksquare$

The projection operator also takes the name *orthogonal projector* in the sense that the error $f - P_N f$ is orthogonal to the polynomial space $\mathbb{P}_N$.

Theorem 3.4. *For any $f \in L_w^2(I)$ and $N \in \mathbb{N}_0$,*

$$\int_I (f - P_N f)\phi w\, dx = (f - P_N f, \phi)_{L_w^2} = 0, \qquad \forall \phi \in \mathbb{P}_N. \tag{3.32}$$

Proof. Let $\phi \in \mathbb{P}_N$ and define $G : \mathbb{R} \to \mathbb{R}$ by

$$G(v) \triangleq \|f - P_N f + v\phi\|_{L_w^2}^2, \qquad v \in \mathbb{R}.$$

From theorem 3.3, $v = 0$ is a minimum of G. Therefore,

$$G'(v) = 2\int_I (f - P_N f)\phi w\, dx + 2v\|\phi\|_{L_w^2}^2$$

should satisfy $G'(0) = 0$. And (3.32) follows directly. $\qquad\blacksquare$

From (3.32) we immediately obtain the Schwarz inequality

$$\|P_N f\|_{L_w^2} \le \|f\|_{L_w^2} \tag{3.33}$$

and the Parseval identity

$$\|f\|_{L_w^2}^2 = \sum_{k=0}^{\infty} \hat{f}_k^2 \|\phi_k\|_{L_w^2}^2. \tag{3.34}$$

3.3.2 Spectral Convergence

The convergence of the orthogonal projection can be stated as follows.

Theorem 3.5. *For any $f \in L_w^2(I)$,*

$$\lim_{N \to \infty} \|f - P_N f\|_{L_w^2} = 0. \tag{3.35}$$

We skip the proof here. When I is bounded, the proof is straightforward; see, for example, [34]. When I is not bounded, the proof is more delicate and we refer readers to [22] for details.

The rate of convergence depends on the regularity of f and the type of orthogonal polynomials $\{\phi_k\}$. There is a large amount of literature devoted to this subject. As a demonstration we present here a result for Legendre polynomials.

Define a *weighted Sobolev space* $H_w^k(I)$, for $k = 0, 1, 2, \ldots$, by

$$H_w^k(I) \triangleq \left\{ v : I \to \mathbb{R} \,\middle|\, \frac{d^m v}{dx^m} \in L_w^2(I), 0 \leq m \leq k \right\}, \tag{3.36}$$

equipped an inner product

$$(u, v)_{H_w^k} \triangleq \sum_{m=0}^k \left(\frac{d^m u}{dx^m}, \frac{d^m v}{dx^m} \right)_{L_w^2} \tag{3.37}$$

and a norm $\|u\|_{H_w^k} = (u, u)_{H_w^k}^{1/2}$.

Let us consider the case of $\bar{I} = [-1, 1]$ with weight function $w(x) = 1$ and Legendre polynomials $\{P_n(x)\}$ (section 3.1.4). The orthogonal projection for any $f(x) \in L_w^2(I)$ is

$$P_N f(x) = \sum_{k=0}^N \hat{f}_k P_k(x), \quad \hat{f}_k = \frac{1}{\|P_k\|_{L_w^2}^2} (f, P_k)_{L_w^2}. \tag{3.38}$$

The following result holds.

Theorem 3.6. *For any $f(x) \in H_w^p[-1, 1]$, $p \geq 0$, there exists a constant C, independent of N, such that*

$$\|f - P_N f\|_{L_w^2[-1,1]} \leq CN^{-p} \|f\|_{H_w^p[-1,1]}. \tag{3.39}$$

Proof. Since the Legendre polynomials satisfy (see (A.25) in appendix A)

$$Q[P_k] = \lambda_k P_k,$$

where

$$Q = \frac{d}{dx}\left((1 - x^2) \frac{d}{dx} \right) = (1 - x^2) \frac{d^2}{dx^2} - 2x \frac{d}{dx}$$

and $\lambda_k = -k(k + 1)$. We then have

$$(f, P_k)_{L_w^2} = \frac{1}{\lambda_k} \int_{-1}^1 Q[P_k] f(x) dx = \frac{1}{\lambda_k} \int_{-1}^1 \left((1 - x^2) P_k'' f - 2x P_k' f \right) dx$$

$$= -\frac{1}{\lambda_k} \int_{-1}^1 \left[((1 - x^2) f)' P_k' + 2x P_k' f \right] dx,$$

where integration by parts has been applied to the first term of the integrand to derive the last equality. Upon simplifying the last expression, we obtain

$$(f, P_k)_{L_w^2} = -\frac{1}{\lambda_k} \int_{-1}^{1} (1 - x^2) f' P_k' dx = \frac{1}{\lambda_k} \int_{-1}^{1} \left((1 - x^2) f'\right)' P_k dx,$$

where integration by parts is again utilized. This implies

$$(f, P_k)_{L_w^2} = \frac{1}{\lambda_k} (Q[f], P_k)_{L_w^2}.$$

By applying the procedure repeatedly for m times, we obtain

$$(f, P_k)_{L_w^2} = \frac{1}{\lambda_k^m} \left(Q^m[f], P_k\right)_{L_w^2}.$$

The projection error can be estimated as

$$
\begin{aligned}
\| f - P_N f \|_{L_w^2}^2 &= \sum_{k=N+1}^{\infty} \hat{f}_k^2 \| P_k \|_{L_w^2}^2 = \sum_{k=N+1}^{\infty} \frac{1}{\| P_k \|_{L_w^2}^2} (f, P_k)_{L_w^2}^2 \\
&= \sum_{k=N+1}^{\infty} \frac{1}{\lambda_k^{2m} \| P_k \|_{L_w^2}^2} (Q^m[f], P_k)_{L_w^2}^2 \\
&\leq \lambda_N^{-2m} \sum_{k=0}^{\infty} \frac{1}{\| P_k \|_{L_w^2}^2} (Q^m[f], P_k)_{L_w^2}^2 \\
&\leq N^{-4m} \| Q^m[f] \|_{L_w^2}^2 \\
&\leq C N^{-4m} \| f \|_{H_w^{2m}}^2,
\end{aligned}
$$

where the last inequality relies on $\| Q^m[f] \|_{L_w^2} \leq C \| f \|_{H_w^{2m}}$, which is a direct consequence of the definitions of $Q^m[f]$ and the norms. By taking $p = 2m$, the theorem is established.

Therefore, the rate of convergence of the Legendre approximation relies on the smoothness of the function f, measured by its differentiability. For a fixed approximation order N, the smoother the function f, the larger the value of p, and the smaller the approximation error. This kind of convergence rate is referred to in the literature as *spectral convergence*. It is in contrast to the traditional finite difference or finite element approximations where the rate of convergence is fixed regardless of the smoothness of the function. An example of spectral convergence is shown in figure 3.2, where the error convergence of the Legendre projections of $|\sin(\pi x)|^3$ and $|x|$ are given. Both functions have finite, but different, smoothness. The convergence rates of the two functions are clearly different in this log-log figure, with $|\sin(\pi x)|^3$ having a faster rate because of its higher differentiability.

If $f(x)$ is analytic, i.e., infinitely smooth, the convergence rate is faster than any algebraic order and we expect

$$\| f - P_N f \|_{L_w^2} \sim C e^{-\alpha N} \| f \|_{L_w^2},$$

where C and α are generic positive constants. Thus, for an analytic function, spectral convergence becomes exponential convergence. An example of exponential

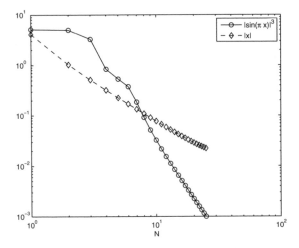

Figure 3.2 Spectral convergence: projection error of $|\sin(\pi x)|^3$ and $|x|$ by Legendre polynomials in $x \in [-1, 1]$. (N is the order of expansion.)

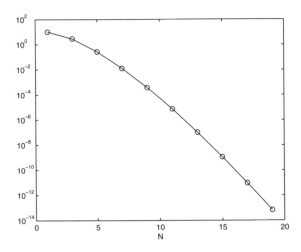

Figure 3.3 Exponential convergence: projection error of $\cos(\pi x)$ by Legendre polynomials in $x \in [-1, 1]$. (N is the order of expansion.)

convergence is shown in figure 3.3, where the projection error of $\cos(\pi x)$ by Legendre polynomials is plotted. Exponential convergence is visible in this kind of semi-log plot.

3.3.3 Gibbs Phenomenon

When the function f is not analytic, the rate of convergence of the polynomial projection is no longer faster than the algebraic rate. In the case of discontinuous functions, the convergence rate deteriorates significantly. For example, consider the

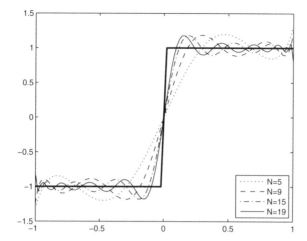

Figure 3.4 Orthogonal series expansion of the sign function by Legendre polynomials. (N is the order of expansion.)

sign function in $(-1, 1)$:

$$\text{sgn}(x) = \begin{cases} 1, & x > 0; \\ -1, & x < 0. \end{cases}$$

The Legendre series expansion is

$$\text{sgn}(x) = \sum_{n=0}^{\infty} \frac{(-1)^n (4n + 3)(2n)!}{2^{2n+1}(n + 1)!n!} P_{2n+1}(x).$$

The partial sums of the series are plotted in figure 3.4 for several values of N. We observe oscillations near the discontinuity, and they do not disappear as N is increased. This is referred to as the *Gibbs phenomenon*. It is a numerical artifact of using globally smooth polynomial basis functions to approximate a discontinuous function. In fact, for this Legendre series expansion, the Gibbs phenomenon has a long-range effect in the sense that it seriously affects the rate of convergence at the endpoints $x = \pm 1$ of the interval. For more a detailed discussion of the Gibbs phenomenon, see [47].

3.4 POLYNOMIAL INTERPOLATION

The goal of polynomial interpolation is to construct a polynomial approximation to a function whose values are known at some discrete points. More precisely, given $m + 1$ pairs of (x_i, y_i), the problem consists of funding a function $G = G(x)$ such that $G(x_i) = y_i$ for $i = 0, \ldots, m$, where y_i are given values and G *interpolates* $\{y_i\}$ at the nodes $\{x_i\}$. In polynomial interpolation, G can be an algebraic polynomial, a trigonometric polynomial, a piecewise polynomial (that is, a local polynomial), a rational polynomial, etc. In the brief introduction here, we focus on global polynomials of algebraic form.

3.4.1 Existence

Let us consider $N + 1$ pairs of (x_i, y_i). The problem is to find a polynomial $Q_M \in \mathbb{P}_M$, called an *interpolating polynomial*, such that

$$Q_M(x_i) = a_M x_i^M + \cdots + a_1 x_i + a_0 = y_i, \qquad i = 0, \ldots, N. \qquad (3.40)$$

The points $\{x_i\}$ are *interpolation nodes*. If $N \neq M$, the problem is over- or underdetermined. If $N = M$, the following results hold.

Theorem 3.7. *Given $N + 1$ distinct points $x_0, \ldots, x_N$ and $N + 1$ corresponding values $y_0, \ldots, y_N$, there exists a unique polynomial $Q_N \in \mathbb{P}_N$ such that $Q_N(x_i) = y_i$ for $i = 0, \ldots, N$.*

Proof. To prove existence, let us use a constructive approach that provides an expression for Q_N. Denoting $\{l_i\}_{i=0}^N$ as a basis for $\mathbb{P}_N$, then $Q_N(x) = \sum_{j=0}^N b_j l_j(x)$ with the property that

$$Q_N(x_i) = \sum_{j=0}^N b_j l_j(x_i) = y_i, \qquad i = 0, \ldots, N.$$

Let us define

$$l_i \in \mathbb{P}_N: \quad l_i(x) = \prod_{\substack{j=0 \\ j \neq i}}^{N+1} \frac{x - x_j}{x_i - x_j}, \qquad i = 0, \ldots, N; \qquad (3.41)$$

then $l_i(x_j) = \delta_{i,j}$ and we obtain $b_i = y_i$.

It is easy to verify that $\{l_i, i = 0, \ldots, N\}$ form a basis for $\mathbb{P}_N$ (left as an exercise). Consequently, the interpolating polynomial exists and has the following form, called the *Lagrange form*,

$$Q_N(x) = \sum_{i=0}^N y_i l_i(x). \qquad (3.42)$$

To prove uniqueness, suppose that another interpolating polynomial $\hat{Q}_M(x)$ of degree $M \leq N$ exists such that $\hat{Q}_M(x_i) = y_i, i = 0, \ldots, N$. Then, the difference polynomial $Q_N - \hat{Q}_M$ vanishes at $N + 1$ distinct points x_i and thus coincides with the null polynomial. Therefore, $\hat{Q}_M = Q_N$.

Another approach also provides a way of constructing the interpolating polynomial. Let

$$Q_N(x_i) = a_N x_i^N + \cdots + a_1 x_i + a_0 = y_i, \qquad i = 0, \ldots, N. \qquad (3.43)$$

This is a system of $N + 1$ equations with $N + 1$ unknowns of the coefficients $a_0, \ldots, a_N$. By letting $\mathbf{a} = (a_0, \ldots, a_N)^T$, $\mathbf{y} = (y_0, \ldots, y_N)^T$, and $\mathbf{A} = (a_{ij}) = (x_i^j)$, system (3.43) can be written as

$$\mathbf{Aa} = \mathbf{y}.$$

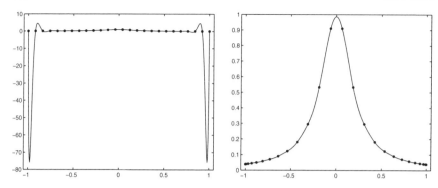

Figure 3.5 Polynomial interpolation of $f(x) = 1/(1+25x^2)$ on $[-1, 1]$, the rescaled Runge function. Left: interpolation on uniformly distributed nodes. Right: interpolation on nonuniform nodes (the zeros of Chebyshev polynomials).

The matrix $\mathbf{A}$ is called a *Vandermonde matrix*, which can be shown to be nonsingular (left as an exercise). Therefore, a unique set of solutions to the coefficients $\mathbf{a}$ exists.

3.4.2 Interpolation Error

Let $f(x), x \in I$, be a given function and let $\Pi_N f(x)$ be its interpolating polynomial of the Nth degree constructed by using the values of $f(x)$ at $N + 1$ distinct points. Then the following result holds.

Theorem 3.8. *Let $x_0, \ldots, x_N$ be $N+1$ distinct nodes and let x be a point inside the interval I. Assume that $f \in C^{N+1}(I_x)$, where I_x is the smallest interval containing the nodes $x_0, \ldots, x_N$ and x. Then the interpolation error at the point x is*

$$E_N(x) = f(x) - \Pi_N f(x) = \frac{f^{N+1}(\xi)}{(N+1)!} q_{N+1}(x), \qquad (3.44)$$

where $\xi \in I_x$ and $q_{N+1} = \prod_{i=0}^{N+1}(x - x_i)$ is the nodal polynomial of degree $N + 1$.

The proof for this standard result is skipped here. (See, for example, [6]).

We should note that a high-degree polynomial interpolation with a set of uniformly distributed nodes is likely to lead to problems, with $q_{N+1}(x)$ behaving rather wildly near the endpoint nodes. This leads to $\Pi_N f(x)$ failing to converge for simple functions such as $f(x) = (1 + x^2)^{-1}$ on $[-5, 5]$, a famous example due to Carl Runge termed the *Runge phenomenon*. To circumvent the difficulty, it is essential to utilize piecewise low-degree polynomial interpolation or high-degree interpolation on nonuniform nodes. In the latter approach, the zeros of orthogonal polynomials provide an excellent choice. This can be seen in figure 3.5, where the Runge function is rescaled to domain $[-1, 1]$ and interpolated by 26 points ($N = 25$). The interpolation on uniformly distributed nodes (on the left in figure 3.5) becomes ill-conditioned and incurs large errors. The interpolation on nonuniformly distributed nodes (on the right in figure 3.5) is stable and accurate.

3.5 ZEROS OF ORTHOGONAL POLYNOMIALS AND QUADRATURE

It is well known that a polynomial of degree N has at most N distinct complex zeros. Although very little can be said about real zeros in general cases, the following result holds.

Theorem 3.9. *Let $\{Q_n(x)\}_{n\in\mathbb{N}}$, $x \in I$, be orthogonal polynomials satisfying orthogonal relation (3.8). Then, for any $n \geq 1$, Q_n has exactly n real distinct zeros in I.*

Proof. First note that $(Q_n, 1)_w = 0$. Therefore, Q_n changes sign in I, hence it has at least one real zero $z_1 \in I$. And $n = 1$ is proved. For $n > 1$, we can find another zero $z_2 \in I$ with $z_2 \neq z_1$ since if Q_n vanishes only at z_1, then the polynomial $(x - z_1)Q_n$ would not change sign in I, which is in contradiction to the relation $((x - z_1), Q_n)_w = 0$, obtained by orthogonality. In a similar fashion, we consider the polynomial $(x - z_1)(x - z_2)Q_n$ and, if $n > 2$, deduce the existence of a third zero, and so on. The procedure ends when all n zeroes are obtained.

There are many properties of the zeros of orthogonal polynomials. For example, one can prove that Q_n and Q_{n-1} do not have common zeros. Moreover, between any two neighboring zeros of Q_{n-1}, there exists one and only one zero of Q_n. Let us recall the following statement.

Theorem 3.10. *Let $\{Q_n\}_{n\in\mathbb{N}}$ be a sequence of orthogonal polynomials in I. Then, for any interval $[a, b] \subset I$, $a < b$, it is possible to find $m \in \mathbb{N}$ such that Q_m has at least one zero in $[a, b]$.*

In other words, this theorem states that the set $J = \bigcup_{n\geq 1} \bigcup_{k=1}^{n} \{z_k^{(n)}\}$ is dense in $\bar{I}$, where $\{z_k^{(n)}\}$ are the zeros of the orthogonal polynomials Q_n. The proof can be found in classical texts on polynomials such as [100].

For example, the zeros of the Legendre polynomials $P_n(x)$, $x \in [-1, 1]$, satisfy

$$-1 \leq -\cos\frac{k - \frac{1}{2}}{n + \frac{1}{2}}\pi \leq z_k^{(n)} \leq -\cos\frac{k}{n + \frac{1}{2}}\pi \leq 1, \qquad 1 \leq k \leq n.$$

The length of the interval between two consecutive zeros is

$$L = -\cos\frac{k}{n + \frac{1}{2}}\pi + \cos\frac{k - \frac{1}{2}}{n + \frac{1}{2}}\pi = 2\sin\frac{2k - \frac{1}{2}}{2n + 1}\pi \sin\frac{\frac{1}{2}}{2n + 1}\pi.$$

For large n, $L \propto n^{-2}$ for $k \approx 1$ or $k \approx n$; and $L \propto n^{-1}$ for moderate values of k. This indicates that the zeros of Legendre polynomials are clustered toward the endpoints of the interval $[-1, 1]$, a feature shared by many orthogonal polynomials.

The nonuniform distribution of the zeros of orthogonal polynomials makes them excellent candidates for polynomial interpolation (see figure 3.5). In addition, they are also excellent for numerical integration.

Let

$$I[f] \triangleq \int_I f(x)w(x)dx \qquad (3.45)$$

and define an *integration formula* with $q \geq 1$ points,

$$U^q[f] \triangleq \sum_{j=1}^{q} f(x^{(j)})w^{(j)}, \qquad (3.46)$$

where $x^{(j)}$ are a set of nodes and $w^{(j)}$ are integration weights, $j = 1, \ldots, q$. The objective is to find a set of $\{x^{(j)}, w^{(j)}\}$ such that $U^q[f] \approx I[f]$ and hopefully $\lim_{q \to \infty} U^q[f] = I[f]$. For example, the well-known trapezoidal rule approximates (3.45) in an interval $[a, b]$ by

$$I[f] \approx \frac{b-a}{2}[f(a) + f(b)],$$

which implies that, in (3.46), $q = 2$, $\{x^{(j)}\} = \{a, b\}$, and $w^{(j)} = \frac{b-a}{2}$, $j = 1, 2$.

Highly accurate integration formulas can be constructed by using orthogonal polynomials. Let $\{Q_n\}_{n \in \mathbb{N}}$ be orthogonal polynomials satisfying (3.8) and let $\{z_k^{(N)}\}_{k=1}^{N}$ be the zeros of Q_N. Let $l_j^{(N-1)}$ be the $(N-1)$th-degree Lagrange polynomials through the nodes $z_j^{(N)}$, $1 \leq j \leq N$, and let $\Pi_{N-1}f(x) = \sum_{j=1}^{N} f(z_j^{(N)})l_j^{(N-1)}$ be the $(N-1)$th-degree interpolation of $f(x)$. Then, the integral (7.12) can be approximated by integrating $\Pi_{N-1}f(x)$,

$$\int_I f(x)w(x)dx \approx \sum_{j=1}^{N} f(z_j^{(N)})w_j^{(N)}, \qquad (3.47)$$

where

$$w_j^{(N)} = \int_I l_j^{(N)}w\,dx, \qquad 1 \leq j \leq N,$$

are the weights. This approximation is obviously exact if $f \in \mathbb{P}_{N-1}$. However, the following result indicates that it is more accurate than this.

Theorem 3.11. *Formula (3.47) is exact; i.e., it becomes an equality if $f(x)$ is any polynomial of degree less than or equal to $2N - 1$ in I.*

Proof. For any $f \in \mathbb{P}_{2N-1}$, let $q = \Pi_{N-1}f \in \mathbb{P}_{N-1}$ be the $(N-1)$th-degree interpolation using $\{z_j^{(N)}\}_{j=1}^{N}$. Then $f - q$ vanishes at $\{z_j^{(N)}\}_{j=1}^{N}$ and can be expressed as $f - q = Q_N(x)r(x)$, where $r(x)$ is a polynomial of degree at most $N - 1$. By orthogonality and using the exactness of (3.47) for polynomials of degree up to $N - 1$, we have

$$\int_I fw\,dx = \int_I qw\,dx + \int_I Q_N r w\,dx = \int_I qw\,dx$$

$$= \sum_{j=1}^{N} q(z_j^{(N)})w_j^{(N)} = \sum_{j=1}^{N} f(z_j^{(N)})w_j^{(N)}.$$

The converse of the theorem is also true; i.e., if formula (3.47) holds for any $f \in \mathbb{P}_{2N-1}$, then the nodes are zeros of Q_N. The degree of the integration formula cannot be improved further. In fact, if one takes $f = Q_N^2 \in \mathbb{P}_{2N}$, the right-hand side of (3.47) vanishes because $Q_N(z_j^{(N)}) = 0$, $j = 1, \ldots, N$, but the left-hand side does not.

3.6 DISCRETE PROJECTION

Here we define a *discrete projection* of a given function $f \in L_w^2(I)$ as

$$I_N f(x) \triangleq \sum_{n=0}^{N} \widetilde{f}_n \phi_n(x), \tag{3.48}$$

where the expansion coefficients

$$\widetilde{f}_n = \frac{1}{\|\phi_n\|_{L_w^2}^2} U^q[f(x)\phi_n(x)] = \frac{1}{\|\phi_n\|_{L_w^2}^2} \sum_{j=1}^{q} f(x^{(j)})\phi_n(x^{(j)})w_j, \qquad 0 \le n \le N. \tag{3.49}$$

When an integration formula U^q is used, the coefficients $\widetilde{f}_n$ are approximations of the coefficients $\hat{f}_n$ (3.30) in the continuous orthogonal projection (3.29). That is,

$$\widetilde{f}_n \approx \hat{f}_n = \int_I f\phi_n w \, dx, \qquad 0 \le n \le N.$$

Note that this definition is slightly more general than what is often used in the literature, in the sense that we did not specify the type and number of nodes used in the integration formula. The only requirement is that the integration formula approximates the corresponding continuous integrals.

In classical spectral methods analysis, the integration formula is typically chosen to be the Gauss quadrature corresponding to the orthogonal polynomials $\{\phi_n(x)\}$ satisfying (3.28). More specifically, let $\{z_j^{(N)}\}_{j=1}^{N+1}$ be the zeros of $\phi_{N+1}(x)$ in $\bar{I}$. Let $\Pi_N f$ be the Lagrange interpolation of $f(x)$ in the form of (3.42),

$$\Pi_N f(x) = \sum_{j=1}^{N+1} f(z_j^{(N)})l_j(x). \tag{3.50}$$

Let us now use the $(N+1)$-point Gauss quadrature to evaluate the coefficients in the discrete expansion (3.48). That is,

$$\widetilde{f}_n = \frac{1}{\|\phi_n\|_{L_w^2}^2} \sum_{j=1}^{N+1} f(z_j^{(N)})\phi_n(z_j^{(N)})w_j^{(N)}, \qquad 0 \le n \le N. \tag{3.51}$$

Then, the following results hold.

Theorem 3.12. *Let $I_N f$ be defined by (3.48), where the coefficients $\{\widetilde{f}_n\}$ are evaluated by $(N+1)$-point Gauss quadrature based on orthogonal polynomial $\{\phi_n\}$, as in (3.51). Let $\Pi_N f$ be the Lagrange interpolation of f through the same set of Gauss nodes, as in (3.50). Then, for any $f \in \mathbb{P}_N$, $\Pi_N f = I_N f$.*

Proof. For $f \in \mathbb{P}_N$, $\Pi_N f = f$ because of the uniqueness of Lagrange interpolation. The coefficients of its discrete expansion are

$$\tilde{f}_n = \frac{1}{\|\phi_n\|_{L_w^2}^2} \sum_{j=1}^{N+1} f(z_j^{(N)}) \phi_n(z_j^{(N)}) w_j^{(N)}$$

$$= \frac{1}{\|\phi_n\|_{L_w^2}^2} \int_I f(x) \phi_n(x) w(x) dx$$

$$= \hat{f}_n, \qquad 0 \leq n \leq N,$$

because the $(N+1)$-point Gauss quadrature is exact for integrating polynomials of degree up to $2N + 1$ and the integrand is in $\mathbb{P}_{2N}$. Therefore, $I_N f = P_N f$, the continuous orthogonal projection of f, which in turn equals f. The proof is established.

The above equivalence of the discrete projection, continuous projection, and Lagrange interpolation does not hold for a general function f because the Gauss quadrature will not be exact. In such a case, $\tilde{f}_n \neq \hat{f}_n$. Hence, $P_N f \neq I_N f$. The difference between the continuous orthogonal projection and the discrete projection is often termed *aliasing error*,

$$A_N f \triangleq P_N f - I_N f. \tag{3.52}$$

By letting $\gamma_n = \|\phi_n\|_{L_w^2}^2$ and realizing that the summation in (3.51) defines a discrete inner product $[f, \phi_n]_w$, we obtain, for all $0 \leq n \leq N$,

$$\tilde{f}_n = \frac{1}{\gamma_n}[f, \phi_n]_w = \frac{1}{\gamma_n}\left[\sum_{j=0}^{\infty} \hat{f}_j \phi_j, \phi_n\right]_w = \frac{1}{\gamma_n}\sum_{j=0}^{\infty} \hat{f}_j [\phi_j, \phi_n]_w$$

$$= \frac{1}{\gamma_n}\left(\sum_{j \leq N}(\phi_j, \phi_n)_{L_w^2} \hat{f}_j + \sum_{j > N}[\phi_j, \phi_n]_w \hat{f}_j\right)$$

$$= \hat{f}_n + \frac{1}{\gamma_n}\sum_{j > N}^{\infty} \hat{f}_j [\phi_j, \phi_n]_w.$$

Therefore,

$$A_N f = \sum_{n=0}^{N}(\hat{f}_n - \tilde{f}_n)\phi_n$$

$$= \sum_{n=0}^{N} \frac{1}{\gamma_n}\sum_{j>N}^{\infty} \hat{f}_j [\phi_j, \phi_n]_w \phi_n = \sum_{j>N}^{\infty}\sum_{n=0}^{N} \frac{1}{\gamma_n}\phi_n [\phi_j, \phi_n]_w \hat{f}_j$$

$$= \sum_{j>N}^{\infty}(I_N \phi_j)\hat{f}_j.$$

Thus, the aliasing error can be seen as the error introduced by using the interpolation of the basis, $I_N \phi_j$, rather than the basis itself to represent the higher expansion modes ($j > N$). The aliasing error stems from the fact that one cannot distinguish between lower and higher basis modes on a finite grid. A general result holds that the aliasing error induced by Gauss points is usually of the same order as that of the projection error. Hence for well-resolved smooth functions, the qualitative behavior of the continuous and the discrete expansions is similar for all practical purposes. We will not pursue a further in-depth discussion of this and instead refer interested readers to the literature, for example, [51].

Chapter Four

Formulation of Stochastic Systems

This chapter is devoted to the general aspects of formulating stochastic equations, i.e., given an established deterministic model for a physical system, how to properly set up a stochastic model to study the effect of uncertainty in the inputs to the system. Prior to any simulation, the key step is to properly characterize the random inputs. More specifically, the goal is to reduce the infinite-dimensional probability space to a finite-dimensional space that is amenable to computing. This is accomplished by parameterizing the probability space by a set of a finite number of random variables. More importantly, it is desirable to require the set of random variables to be mutually independent. We remark that the independence requirement is very much a concern from a practical point of view, for most, if not all, available numerical techniques require independence. This is not as strong a requirement from a theoretical point of view. Although there exist some techniques to loosen it, in this book we shall continue to employ this widely adopted requirement.

To summarize, the critical step in formulating a stochastic system is to properly *characterize the probability space defined by the random inputs by a set of a finite number of mutually independent random variables.* In many cases such a characterization procedure cannot be done exactly and will induce approximation errors.

4.1 INPUT PARAMETERIZATION: RANDOM PARAMETERS

When the random inputs to a system are the system parameters, the parameterization procedure is straightforward, for the inputs are already in the form of parameters. The more important issue is then to identify the *independent* parameters in the set. The problem can be stated as follows.

> Let $Y = (Y_1, \ldots, Y_n)$, $n > 1$, be the system parameters with a prescribed distribution function $F_Y(y) = P(Y \leq y)$, $y \in \mathbb{R}^n$, and find a set of mutually independent random variables $Z = (Z_1, \ldots, Z_d) \in \mathbb{R}^d$, where $1 \leq d \leq n$, such that $Y = T(Z)$ for a suitable transformation function T.

Let us use a simple example to illustrate the idea. Consider an ordinary differential equation with two random parameters,

$$\frac{du}{dt}(t, \omega) = -\alpha(\omega)u, \qquad u(0, \omega) = \beta(\omega), \tag{4.1}$$

where the rate constant α and the initial condition β are assumed to be random. Thus, the input random variables are $Y(\omega) = (\alpha, \beta) \in \mathbb{R}^2$.

If α and β are mutually independent, then we simply let $Z(\omega) = Y(\omega)$. The solution

$$u(t, \omega) : [0, T] \times \Omega \to \mathbb{R}$$

can now be expressed as

$$u(t, Z) : [0, T] \times \mathbb{R}^2 \to \mathbb{R},$$

which has one time dimension and two random dimensions.

If α and β are not independent of each other, it implies that there exists a function f such that

$$f(\alpha, \beta) = 0.$$

Then it is possible to find a random variable $Z(\omega)$ to parameterize the relation such that

$$\alpha(\omega) = a(Z(\omega)), \quad \beta(\omega) = b(Z(\omega)),$$

and $f(a, b) = 0$. Or, equivalently, the dependence between α and β implies that there exists a function g such that

$$\beta = g(\alpha).$$

Then we can let $Z(\omega) = \alpha(\omega)$ and $\beta(\omega) = g(Z(\omega))$. Which approach is more convenient in practice is problem-dependent. Nevertheless, the random inputs via α and β can now be expressed via a single random variable Z, and the solution becomes

$$u(t, Z) : [0, T] \times \mathbb{R} \to \mathbb{R},$$

which now has only one random dimension.

In practice, when there are many parameters in a system, finding the exact form of the functional dependence among all the parameters can be a challenging (and unnecessary) task. This is especially true when the only available information is the (joint) probability distributions of the parameters. The goal now is to transform the parameters to a set of independent random parameters by using their distribution functions.

4.1.1 Gaussian Parameters

Since the first two moments, mean and covariance, can completely characterize Gaussian distribution, the parameterization problem can be solved in a straightforward manner.

Let $Y = (Y_1, \ldots, Y_n)$ be a random vector with a Gaussian distribution of $\mathcal{N}(0, \mathbf{C})$, where $\mathbf{C} \in \mathbb{R}^{n \times n}$ is the covariance matrix and the expectation is assumed to be zero (without loss of generality). Let $Z \sim \mathcal{N}(0, \mathbf{I})$, where $\mathbf{I}$ is the $n \times n$ identity matrix, be a uncorrelated Gaussian vector of size n. Thus, the components of Z are mutually independent. Let $\mathbf{A}$ be an $n \times n$ matrix; then by theorem 2.16,

$\mathbf{A}Z \sim \mathcal{N}(0, \mathbf{A}\mathbf{A}^T)$. Therefore, if one finds a matrix $\mathbf{A}$ such that $\mathbf{A}\mathbf{A}^T = \mathbf{C}$, then $Y = \mathbf{A}Z$ will have the given distribution $\mathcal{N}(0, \mathbf{C})$. This result is a special case of a more general theorem by Anderson [4].

Since $\mathbf{C}$ is real and symmetric, solving the problem $\mathbf{A}\mathbf{A}^T = \mathbf{C}$ can be readily done via, for example, Cholesky's decomposition, where $\mathbf{A}$ takes the form of a lower-triangular matrix:

$$
\begin{aligned}
a_{i1} &= c_{i1}/\sqrt{c_{11}}, & 1 \le i \le n, \\
a_{ii} &= \sqrt{c_{ii} - \sum_{k=1}^{i-1} a_{ik}^2}, & 1 < i \le n, \\
a_{ij} &= \left(c_{ij} - \sum_{k=1}^{j-1} a_{ik} a_{jk} \right) / a_{jj}, & 1 < j < i \le n, \\
a_{ij} &= 0, & i < j \le n,
\end{aligned}
\tag{4.2}
$$

where a_{ij} and c_{ij}, $1 \le i, j \le n$, are the entries for the matrices $\mathbf{A}$ and $\mathbf{C}$, respectively.

4.1.2 Non-Gaussian Parameters

When the system parameters have a non-Gaussian distribution, the parameterization problem is distinctly more difficult. However, there exists a remarkably simple transformation due to Rosenblatt [88] that can accomplish the goal. Let $Y = (Y_1, \ldots, Y_n)$ be a random vector with a (non-Gaussian) distribution function $F_Y(y) = P(Y \le y)$ and let $z = (z_1, \ldots, z_n) = Ty = T(y_1, \ldots, y_n)$ be a transformation defined as

$$
\begin{aligned}
z_1 &= P(Y_1 \le y_1) = F_1(y_1), \\
z_2 &= P(Y_2 \le y_2 | Y_1 = y_1) = F_2(y_2 | y_1), \\
&\cdots, \\
z_n &= P(Y_n \le y_n | Y_{n-1} = y_{n-1}, \ldots, Y_1 = y_1) = F_n(y_n | y_{n-1}, \ldots, y_1).
\end{aligned}
\tag{4.3}
$$

It can then be shown that

$$
\begin{aligned}
P(Z_i &\le z_i; i = 1, \ldots, n) \\
&= \int_{\{Z|Z_i \le z_i\}} \cdots \int d_{y_n} F_n(y_n | y_{n-1}, \ldots, y_1) \cdots d_{y_1} F_1(y_1) \\
&= \int_0^{z_n} \cdots \int_0^{z_1} dz_1 \cdots dz_n = \prod_{i=1}^n z_i,
\end{aligned}
$$

where $0 \le z_i \le 1$, $i = 1, \ldots, n$. Hence $Z = (Z_1, \ldots, Z_n)$ are independent and identically distributed (i.i.d.) random variables with uniform distribution in $[0, 1]^n$.

Though mathematically simple and powerful, Rosenblatt transformation is not easy to carry out in practice, for it relies on the conditional probability distributions among the random parameters. Such information is rarely known completely. And even if it is known, it is rarely given in explicit formulas. In practice, some kinds of numerical approximations of Rosenblatt transformation are required. This remains an understudied topic and is beyond the scope of this book.

4.2 INPUT PARAMETERIZATION: RANDOM PROCESSES AND DIMENSION REDUCTION

In many cases, the random inputs are stochastic processes. For example, the inputs could be a time-dependent random forcing term that is a stochastic process in time, or an uncertain material property, e.g., conductivity, that is a stochastic process in space. The parameterization problem can be stated as follows.

> Let $(Y_t, t \in T)$ be a stochastic process that models the random inputs, where t is the index belonging to an index set T, and find a suitable transformation function R such that $Y_t = R(Z)$, where $Z = (Z_1, \ldots, Z_d)$, $d \geq 1$, are mutually independent.

Note that the index set T can be in either the space domain or the time domain and is usually an infinite-dimensional object. Since we require d to be a finite integer, the transformation cannot be exact. Therefore, $Y_t \approx R(Z)$ in a proper norm or metric, and the accuracy of the approximation will be problem-dependent.

A straightforward approach is to consider the finite-dimensional version of Y_t instead of Y_t directly. This requires one to first discretize the index domain T into a set of finite indices and then study the process

$$(Y_{t_1}, \ldots, Y_{t_n}), \quad t_1, \ldots, t_n \in T,$$

which is now a finite-dimensional random vector. The parameterization techniques for random parameters from the previous section can now be readily applied, e.g., Rosenblatt transformation.

The discretization of the Y_t into its finite-dimensional version is obviously an approximation. The finer the discretization, the better the approximation. However, this is not desired because a finer discretization leads to a larger dimension of n and can significantly increase the computational burden. Some kinds of dimension reduction techniques are required to keep the dimension as low as possible while maintaining a satisfactory approximation accuracy.

4.2.1 Karhunen-Loeve Expansion

The Karhunen-Loeve (KL) expansion (cf. [73], for example) is one of the most widely used techniques for dimension reduction in representing random processes.

Let $\mu_Y(t)$ be the mean of the input process Y_t and let $C(t, s) = \text{cov}(Y_t, Y_s)$ be its covariance function. The Karhunen-Loeve expansion of Y_t is

$$Y_t(\omega) = \mu_Y(t) + \sum_{i=1}^{\infty} \sqrt{\lambda_i} \psi_i(t) Y_i(\omega), \tag{4.4}$$

where ψ_i are the orthogonal eigenfunctions and λ_i are the corresponding eigenvalues of the eigenvalue problem

$$\int_T C(t, s) \psi_i(s) ds = \lambda_i \psi_i(t), \quad t \in T, \tag{4.5}$$

and $\{Y_i(\omega)\}$ are mutually uncorrelated random variables satisfying

$$\mathbb{E}[Y_i] = 0, \quad \mathbb{E}[Y_i Y_j] = \delta_{ij}, \tag{4.6}$$

and defined by

$$Y_i(\omega) = \frac{1}{\sqrt{\lambda_i}} \int_T (Y_t(\omega) - \mu_Y(t))\psi_i(t)dt, \quad \forall i. \tag{4.7}$$

The Karhunen-Loeve expansion, in the form of the equality (4.4), is of little use because it is an infinite series. In practice, one adopts a finite series expansion, e.g.,

$$Y_t(\omega) \approx \mu_Y(t) + \sum_{i=1}^{d} \sqrt{\lambda_i}\psi_i(t)Y_i(\omega), \quad d \geq 1. \tag{4.8}$$

The natural question to ask is when to truncate the series. That is, how to choose d so that the approximation accuracy is satisfactory. The answer to the question is closely related to an important property of the Karhunen-Loeve expansion—decay of the eigenvalues λ_i as index i increases. Here we illustrate the property with the following examples.

Example 4.1 (Exponential covariance function). Let $C(t,s) = \exp(-|t-s|/a)$, where $a > 0$ is the correlation length, and let $t \in T = [-b, b]$ be in a bounded domain with length $2b$. Then the eigenvalue problem (4.5) can be solved analytically [109]. The eigenvalues are

$$\lambda_i = \begin{cases} \frac{2a}{1+a^2w_i^2}, & \text{if } i \text{ is even,} \\ \frac{2a}{1+a^2v_i^2}, & \text{if } i \text{ is odd,} \end{cases} \tag{4.9}$$

and the corresponding eigenfunctions are

$$\psi_i(t) = \begin{cases} \sin(w_i t) \big/ \sqrt{b - \frac{\sin(2w_i b)}{2w_i}}, & \text{if } i \text{ is even,} \\ \cos(v_i t) \big/ \sqrt{b + \frac{\sin(2v_i b)}{2v_i}}, & \text{if } i \text{ is odd,} \end{cases} \tag{4.10}$$

where w_i and v_i are the solutions of the transcendental equations

$$\begin{cases} aw + \tan(wb) = 0, & \text{for even } i, \\ 1 - av\tan(vb) = 0, & \text{for odd } i. \end{cases}$$

In figure 4.1, the first four eigenfunctions are shown for the exponential covariance function in $[-1, 1]$. It is obvious that the higher modes (the eigenfunctions with larger index i) have a finer structure compared to the lower modes. The eigenvalues are shown in figure 4.2 for several different correlation lengths a. It can be seen that the eigenvalues decay, and the decay rate is larger when the correlation length is longer. When the correlation length is very small, e.g., $a = 0.01$, the decay of the eigenvalues is barely visible.

Example 4.2 (Uncorrelated process). The limit of diminishing correlation length is the zero correlation case, when the covariance function takes the form of a delta function, $C(t,s) = \delta(t-s)$. It is easy to see from (4.5) that now any orthogonal functions can be the eigenfunctions and the eigenvalues are a constant, i.e., $\lambda_i = 1$, $\forall i$. In this case, there will not be any decay of the eigenvalues.

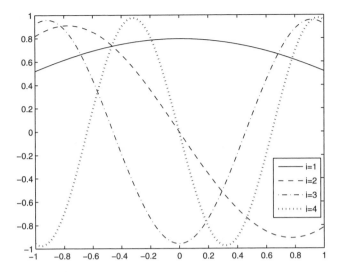

Figure 4.1 The first four eigenfunctions of the exponential covariance function.

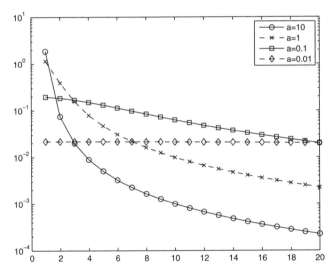

Figure 4.2 The first 20 eigenvalues of the exponential covariance function with different correlation lengths a.

Example 4.3 (Fully correlated process). The other limit is when $C(t, s) = 1$, which implies an infinite correlation length and that the process Y_t is fully correlated. This is the rather trivial case where the process depends on just one random variable. It is straightforward from (4.5) to show that there exists one nonzero eigenvalue corresponding to a constant eigenfunction and that the rest of the eigenvalues are zero.

The aforementioned examples illustrate an important property of the Karhunen-Loeve expansion: for a given covariance function, *the decay rate of the eigenvalues depends inversely on the correlation length.* Long correlation length implies that the process is strongly correlated and results in a fast decay of the eigenvalues. The limit of this, infinitely long correlation length, is the fully correlated case where the eigenvalues decay to zero immediately. Conversely, a weakly correlated process has short correlation length and results in a slow decay of the eigenvalues. The limit of this, the uncorrelated process with zero correlation length, has no eigenvalue decay.

The decay rate of the eigenvalues provides a guideline for truncating the infinite KL series (4.4) into the finite KL series (4.8). The common approach for truncation is to examine the decay of λ_i and keep the first d eigenvalues so that the contribution of the rest of the eigenvalues is negligible. How much is considered negligible, usually given in terms of a small percentage as a cutoff criterion, is specified on a problem-dependent basis. Naturally, for a given cutoff criterion, a strongly correlated process allows a finite KL expansion with a fewer number of terms than a weakly correlated process does.

There are many more properties regarding the KL expansion. For example, the error of a finite-term expansion is optimal in terms of the mean-square error. We will not devote further discussion to this and refer interested readers to references such as [93].

4.2.2 Gaussian Processes

The truncated Karhunen-Loeve series (4.8) provides a way to approximate a random process by a function (a series) involving a finite number of random variables. The set of random variables $Y_i(\omega)$ are uncorrelated, as in (4.6). For Gaussian random variables, uncorrelation and independence are equivalent. Furthermore, linear combinations of Gaussian random variables remain Gaussian-distributed. Therefore, if $Y_t(\omega)$ is a Gaussian process, then the random variables Y_i in (4.4) and (4.8) are independent Gaussian random variables. Hence (4.8) provides a natural way to parameterize a Gaussian process by a finite number of independent Gaussian random variables.

4.2.3 Non-Gaussian Processes

When the input random processes are non-Gaussian, their parameterization and dimension reduction are significantly more challenging. The main problem is that, for non-Gaussian distributions, uncorrelation of the random variables Y_i in (4.4) does not imply independence. Hence the Karhunen-Loeve expansion does not provide a way of parameterization with independent variables. In many practical computations, one often still uses the KL expansion for an input process and then further assumes that the Y_i are independent. Though this is not a rigorous approach, at the moment there are not many practical methods for achieving the parameterization procedure. We will not engage in further discussion of this because it remains an active (and open) research topic.

4.3 FORMULATION OF STOCHASTIC SYSTEMS

We now illustrate the main steps in formulating a stochastic system by taking into account random inputs to a well-established deterministic system. We choose partial differential equations (PDEs) as a basic model, although the concept and procedure are not restricted to PDEs.

Let us consider a system of PDEs defined in a spatial domain $D \subset \mathbb{R}^\ell$, $\ell = 1, 2, 3$, and a time domain $[0, T]$ with $T > 0$,

$$\begin{cases} u_t(x, t, \omega) = \mathcal{L}(u), & D \times (0, T] \times \Omega, \\ \mathcal{B}(u) = 0, & \partial D \times [0, T] \times \Omega, \\ u = u_0, & D \times \{t = 0\} \times \Omega, \end{cases} \quad (4.11)$$

where $\mathcal{L}$ is a (nonlinear) differential operator, $\mathcal{B}$ is the boundary condition operator, u_0 is the initial condition, and $\omega \in \Omega$ denotes the random inputs of the system in a properly defined probability space $(\Omega, \mathcal{F}, P)$. Note in general that it is not important, nor is it relevant, to identify precisely the probability space. The solution is therefore a random quantity,

$$u(x, t, \omega) : \bar{D} \times [0, T] \times \Omega \to \mathbb{R}^{n_u}, \quad (4.12)$$

where $n_u \geq 1$ is the dimension of u.

The random inputs to (4.11) can take the form of random parameters and random processes. Let us assume that they can all be properly parameterized by a set of independent random variables using the techniques discussed in the previous two sections. Let $Z = (Z_1, \ldots, Z_d) \in \mathbb{R}^d$, $d \geq 1$, be the set of independent random variables characterizing the random inputs. We can then rewrite system (4.11) as

$$\begin{cases} u_t(x, t, Z) = \mathcal{L}(u), & D \times (0, T] \times \mathbb{R}^d, \\ \mathcal{B}(u) = 0, & \partial D \times [0, T] \times \mathbb{R}^d, \\ u = u_0, & D \times \{t = 0\} \times \mathbb{R}^d. \end{cases} \quad (4.13)$$

The solution is now

$$u(x, t, Z) : \bar{D} \times [0, T] \times \mathbb{R}^d \to \mathbb{R}^{n_u}. \quad (4.14)$$

The fundamental assumption we make is that (4.11) is a well-posed system P-almost surely in Ω. Loosely and intuitively speaking, this means that if one generates an ensemble of (4.13) by generating a collection of realizations of the random variables Z, then each realization is well posed in its corresponding deterministic sense.

Example 4.4. Consider the same example of the ordinary differential equation (ODE) (4.1)

$$\frac{du}{dt}(t, \omega) = -\alpha(\omega)u, \quad u(0, \omega) = \beta(\omega). \quad (4.15)$$

If the input random variables α and β are independent, then we let $Z = (Z_1, Z_2) = (\alpha, \beta)$ and rewrite the problem as

$$\frac{du}{dt}(t, \omega) = -Z_1 u, \quad u(0, \omega) = Z_2. \tag{4.16}$$

If, however, α and β are dependent, then, as discussed in section 4.1, we can let $Z = \alpha$ and there exists a function such that $\beta = g(\alpha)$. The system can be rewritten as

$$\frac{du}{dt}(t, \omega) = -Zu, \quad u(0, \omega) = g(Z). \tag{4.17}$$

Example 4.5 (Stochastic diffusion equation). Consider a one-dimensional stochastic elliptic equation

$$\begin{cases} \nabla \cdot (\kappa(x, \omega)\nabla u) = f(x, \omega), & x \in (-1, 1), \\ u(-1, \omega) = u_\ell(\omega), & u(1, \omega) = u_r(\omega), \end{cases} \tag{4.18}$$

where the diffusivity field κ and the source term f are assumed to be random fields and u_ℓ and u_r are random variables. For simplicity of exposition, only the Dirichlet boundary condition is considered. Let us assume that the diffusivity field κ can be parameterized by a truncated KL expansion (4.8) with d_κ terms, i.e.,

$$\kappa(x, \omega) \approx \widetilde{\kappa}(x, Z^\kappa) = \mu_\kappa(x) + \sum_{i=1}^{d_\kappa} \hat{\kappa}_i(x) Z_i^\kappa(\omega),$$

where the functions $\hat{\kappa}_i(x)$ are determined by the eigenvalues and eigenfunctions of the covariance function of $\kappa(x, \omega)$, and $Z_i^\kappa(\omega)$ are mutually independent. Similarly, let $f(x, \omega)$ be parameterized as

$$f(x, \omega) \approx \widetilde{f}(x, Z^f) = \mu_f(x) + \sum_{i=1}^{d_f} \hat{f}_i(x) Z_i^f(\omega),$$

with d_f terms and mutually independent $Z_i^f(\omega)$. Let us assume κ and f are independent of each other and also independent of u_ℓ and u_r. Then let

$$Z = (Z_1, \dots, Z_d) = (Z_1^\kappa, \dots, Z_{d_\kappa}^\kappa, Z_1^f, \dots, Z_{d_f}^f, u_\ell, u_r),$$

where $d = d_\kappa + d_f + 2$, and we can write the elliptic problem as

$$\begin{cases} \nabla \cdot (\widetilde{\kappa}(x, Z)\nabla u) = \widetilde{f}(x, Z), & x \in (-1, 1), \\ u(-1, Z) = Z_{d-1}, & u(1, Z) = Z_d. \end{cases} \tag{4.19}$$

The solution is $u(x, Z) : [-1, 1] \times \mathbb{R}^d \to \mathbb{R}$.

4.4 TRADITIONAL NUMERICAL METHODS

Here we briefly review the traditional methods for solving practical systems with random inputs. For the purpose of illustration, we use the simple stochastic ODE in example 4.4 as an example.

The exact solution to (4.15) is

$$u(t, \omega) = \beta(\omega)e^{-\alpha(\omega)t}. \tag{4.20}$$

When the distribution function of α and β is known, i.e., $F_{\alpha\beta}(a, b) = P(\alpha \leq a, \beta \leq b)$, the statistical moments of the solution can be evaluated. If $\alpha(\omega)$ and $\beta(\omega)$ are independent, i.e., $F_{\alpha\beta}(a, b) = F_\alpha(a)F_\beta(b)$, then

$$\mathbb{E}[u(t, \omega)] = \mathbb{E}[\beta]\mathbb{E}\left[e^{-\alpha t}\right].$$

For example, if we further assume $\beta(\omega) \equiv 1$, that is, the initial condition is not random, and $\alpha(\omega) \sim \mathcal{N}(0, 1)$ is a standard Gaussian random variable with zero mean and unit variance, then

$$\mathbb{E}[u] = \frac{1}{\sqrt{2\pi}} \int e^{-at} e^{-a^2/2} da = e^{t^2/2}$$

and

$$\sigma_u^2 = \mathbb{E}\left[u^2\right] - (\mathbb{E}[u])^2 = e^{2t^2} - e^{t^2}.$$

4.4.1 Monte Carlo Sampling

Monte Carlo sampling (MCS) is a statistical sampling method that was popularized by physicists from Los Alamos National Laboratory in the United States in the 1940s. The general procedure for (4.15) is as follows.

1. Generate identically and independently distributed random numbers $Z^{(i)} = (\alpha^{(i)}, \beta^{(i)})$, $i = 1, \ldots, M$, according to the distribution of $F_{\alpha\beta}(a, b)$. Note once again that the dependence structure of α and β is required to be known.
2. For each $i = 1, \ldots, M$, solve the governing equation (4.15) and obtain $u^{(i)}(t) \triangleq u(t, Z^{(i)})$.
3. Estimate the required solution statistics. For example, the solution mean can be estimated as

$$\bar{u}(t) = \frac{1}{M} \sum_{i=1}^{M} u(t, Z^{(i)}) \approx \mathbb{E}[u]. \tag{4.21}$$

Other solution statistics can be estimated via proper schemes from the solution ensemble $\{u^{(i)}\}$. It is obvious that steps 1 and 3 are preprocessing and postprocessing steps, respectively. Only step 2 requires solution of the original problem, and it involves repetitive simulations of the deterministic counterpart of the problem.

Error estimate of MCS follows immediately from the Central Limit Theorem (theorem 2.26). Since $\{u(t, Z^{(i)})\}$ are independent and identically distributed random variables, the distribution function of $\bar{u}(t)$ converges, in the limit of $M \to \infty$, to a Gaussian distribution $\mathcal{N}(\mathbb{E}[u](t), \sigma_u^2(t)/M)$, whose standard deviation is $M^{-1/2}\sigma_u(t)$, where σ_u is the standard deviation of the exact solution. Hence the widely adopted concept that the error convergence rate of MCS is inversely proportional to the square root of the number of realizations.

It is obvious that the MCS procedure can be easily extended to a general system such as (4.13). The only requirement is that one needs a well-established solver to solve the corresponding deterministic system. Although very powerful and flexible, the convergence rate of MCS, $O(M^{-1/2})$, is relatively slow. Generally speaking, if a one-digit increase in solution accuracy of the statistics is required, one needs to run roughly 100 times more simulations and thus increase the computational effort by 100 times. For large and complex systems where the solution of a single deterministic realization is time-consuming, this poses a tremendous numerical challenge. On the other hand, a remarkable advantage of MCS lies in the fact that the $O(M^{-1/2})$ convergence rate is independent of the total number of input random variables. That is, the convergence rate is independent of the dimension of the random space. This turns out to be an extremely useful property that virtually no other methods possess.

4.4.2 Moment Equation Approach

The objective of the moment equation approach is to compute the moments of the solution directly because in many cases the moments of the solution are what one needs.

Let $\mu(t) = \mathbb{E}[u]$; then by taking the expectation of both sides of (4.15) we obtain

$$\frac{d\mu}{dt}(t) = -\mathbb{E}[\alpha u], \qquad \mu(0) = \mathbb{E}[\beta].$$

This requires a knowledge of $\mathbb{E}[\alpha u]$, which is not known. We then attempt to derive an equation for this new quantity. This is done by multiplying (4.15) by α and then taking the expectation,

$$\frac{d}{dt}\mathbb{E}[\alpha u](t) = -\mathbb{E}[\alpha^2 u], \qquad \mathbb{E}[\alpha u](0) = \mathbb{E}[\alpha \beta].$$

A new quantity $\mathbb{E}[\alpha^2 u]$ appears. If we attempt a similar approach by multiplying the original system by α^2 and then taking the expectation, the equation for the new quantity is

$$\frac{d}{dt}\mathbb{E}[\alpha^2 u](t) = -\mathbb{E}[\alpha^3 u], \qquad \mathbb{E}[\alpha^2 u](0) = \mathbb{E}[\alpha^2 \beta],$$

which now requires yet another new quantity, $\mathbb{E}[\alpha^3 u]$. In general, if we define $\widetilde{\mu}_k(t) = \mathbb{E}[\alpha^k u]$ for $k = 0, 1, \ldots$, then we obtain

$$\frac{d}{dt}\widetilde{\mu}_k(t) = -\widetilde{\mu}_{k+1}, \qquad \widetilde{\mu}_k(0) = \mathbb{E}[\alpha^k \beta].$$

The system of equations thus cannot be closed, as it keeps introducing new variables that are not included in the existing system. This is the well-known *closure problem*. The typical approach to remedy the difficulty is to assume, for a fixed k, that the higher-order term is a function of the lower-order ones, that is,

$$\widetilde{\mu}_{k+1} = g(\widetilde{\mu}_0, \ldots, \widetilde{\mu}_k),$$

where the form of the function g is determined on a problem-dependent basis with (hopefully) sufficient justification. There exists, to date, no effective general strategy for solving the closure problem. Also, the error caused by most, if not all, closure techniques is not easy to quantify.

4.4.3 Perturbation Method

In the perturbation method, the random part of the solution is treated as a perturbation. The fundamental assumption is that such a perturbation is *small*. And this is typically achieved by requiring the standard deviation of the solution to be small. To illustrate the idea, let us again use the simple ODE example (4.15). For ease of exposition, let us further assume that the mean value of α is zero, i.e., $\mathbb{E}[\alpha] = 0$, and that the initial condition is a fixed value. That is,

$$\frac{du}{dt}(t, \omega) = -\alpha(\omega)u, \qquad u(0, \omega) = \beta.$$

The perturbation method can be applied when the variation of α is small, that is, $\epsilon = O(\alpha(\omega)) \sim \sigma_\alpha \ll 1$. If this is the case, we seek to expand the solution as a power series of α,

$$u(t, \omega) = u_0(t) + \alpha(\omega)u_1(t) + \alpha^2(\omega)u_2(t) + \cdots, \qquad (4.22)$$

where the coefficients $u_0, u_1, \ldots$ are supposed to be of same order of magnitude. After substituting the expression into the governing equation (4.1), we obtain

$$\frac{du_0}{dt} + \alpha\frac{du_1}{dt} + \alpha^2\frac{du_2}{dt} + \cdots = -\alpha(u_0 + \alpha u_1 + \alpha^2 2u_2 + \cdots).$$

Since $O(\alpha^k) = \epsilon^k$ becomes increasingly small as k increases, we match the terms in the equation according to the power of α.

$$O(1): \qquad \frac{du_0}{dt} = 0,$$

$$O(\epsilon): \qquad \frac{du_1}{dt} = -u_0,$$

$$O(\epsilon^2): \qquad \frac{du_2}{dt} = -u_1,$$

$$\cdots \qquad \cdots$$

$$O(\epsilon^k): \qquad \frac{du_k}{dt} = -u_{k-1},$$

Similar expansion and term matching of the initial condition result in the initial conditions for the coefficients,

$$u_0(0) = \beta, \qquad u_k(0) = 0, \quad k > 1.$$

It is then easy to solve the system recursively and obtain

$$u_0(t) = \beta, \qquad u_1 = -\beta t, \qquad \ldots, \qquad u_k = \beta(-1)^k\frac{t^k}{k!}.$$

The power series then gives us the solution

$$u(t, \omega) = \sum_{k=0}^{\infty} \beta\frac{-t^k}{k!}\alpha^k(\omega), \qquad (4.23)$$

which is the infinite power series of the exact solution $u_{exact}(t, \omega) = \beta\exp(-\alpha t)$.

Even though it seems that the perturbation solution can recover the exact solution in terms of its power series, the requirement for α to be small is still needed. This is because in practice one can use only a finite-term series, which can provide a good approximation *only* when α is small. (Note from (4.23) that when α is $O(1)$ or bigger, the remainder of a finite-term series is always dominant.)

Derivation of the equations for each terms is done by matching the orders. The procedure cannot be easily automated, except for very simple problems such as the example here. For practical systems, one usually stops the procedure at the second-order terms because of the complexity of the derivation. For first- or second-order approximations to be satisfactory, an even stronger requirement for smallness is needed. However, a distinct feature of perturbation methods is that, once derived, the systems of equations are almost always decoupled and can be solved recursively.

Chapter Five

Generalized Polynomial Chaos

This chapter is devoted to the fundamental aspects of generalized polynomial chaos (gPC). The material is largely based on the work described in [120]. However, the exposition here is quite different from the original one in [120], for better understanding of the material. It should also be noted that here we focus on the gPC expansion based on globally smooth orthogonal polynomials, in an effort to illustrate the basic ideas, and leave other types of gPC expansion (e.g., those based on piecewise polynomials) as research topics outside the scope of this book.

5.1 DEFINITION IN SINGLE RANDOM VARIABLES

Let Z be a random variable with a distribution function $F_Z(z) = P(Z \leq z)$ and finite moments

$$\mathbb{E}\left(|Z|^{2m}\right) = \int |z|^{2m} d F_Z(z) < \infty, \qquad m \in \mathcal{N}, \tag{5.1}$$

where $\mathcal{N} = \mathbb{N}_0 = \{0, 1, \dots\}$ or $\mathcal{N} = \{0, 1, \dots, N\}$ and for a finite nonnegative integer N is an index set. The *generalized polynomial chaos* basis functions are the orthogonal polynomial functions satisfying

$$\mathbb{E}[\Phi_m(Z)\Phi_n(Z)] = \gamma_n \delta_{mn}, \qquad m, n \in \mathcal{N}, \tag{5.2}$$

where

$$\gamma_n = \mathbb{E}\left[\Phi_n^2(Z)\right], \qquad n \in \mathcal{N}, \tag{5.3}$$

are the normalization factors.

If Z is continuous, then its probability density function (PDF) exists such that $d F_Z(z) = \rho(z)dz$ and the orthogonality can be written as

$$\mathbb{E}[\Phi_m(Z)\Phi_n(Z)] = \int \Phi_m(z)\Phi_n(z)\rho(z)dz = \gamma_n \delta_{mn}, \qquad m, n \in \mathcal{N}. \tag{5.4}$$

Similarly, when Z is discrete, the orthogonality can be written as

$$\mathbb{E}[\Phi_m(Z)\Phi_n(Z)] = \sum_i \Phi_m(z_i)\Phi_n(z_i)\rho_i = \gamma_n \delta_{mn}, \qquad m, n \in \mathcal{N}. \tag{5.5}$$

With a slight abuse of notation, hereafter we will use

$$\mathbb{E}[f(Z)] = \int f(z)d F_Z(z)$$

to include both the continuous case and the discrete case.

Obviously, $\{\Phi_m(z)\}$ are orthogonal polynomials of $z \in \mathbb{R}$ with the weight function $\rho(z)$, which is the probability function of the random variable Z. This establishes a correspondence between the distribution of the random variable Z and the type of orthogonal polynomials of its gPC basis.

Example 5.1 (Hermite polynomial chaos). Let $Z \sim \mathcal{N}(0, 1)$ be a standard Gaussian random variable with zero mean and unit variance. Its PDF is

$$\rho(z) = \frac{1}{\sqrt{2\pi}} e^{-z^2/2}.$$

The orthogonality (5.2) then defines the Hermite orthogonal polynomials $\{H_m(Z)\}$ as in (3.19). Therefore, we employ the Hermite polynomials as the basis functions,

$$H_0(Z) = 1, \quad H_1(Z) = Z, \quad H_2(Z) = Z^2 - 1, \quad H_3(Z) = Z^3 - 3Z, \quad \dots.$$

This is the classical Wiener-Hermite polynomial chaos basis ([45]).

Example 5.2 (Legendre polynomial chaos). Let $Z \sim \mathcal{U}(-1, 1)$ be a random variable uniformly distributed in $(-1, 1)$. Its PDF is $\rho(z) = 1/2$ and is a constant. The orthogonality (5.2) then defines the Legendre orthogonal polynomials (3.16), with

$$L_0(Z) = 1, \quad L_1(Z) = Z, \quad L_2(Z) = \frac{3}{2}Z^2 - \frac{1}{2}, \quad \dots.$$

Example 5.3 (Jacobi polynomial chaos). Let Z be a random variable of beta distribution in $(-1, 1)$ with PDF

$$\rho(z) \propto (1 - z)^\alpha (1 + z)^\beta, \qquad \alpha, \beta > 0,$$

whose precise definition is in (A.21). The orthogonality (5.2) then defines the Jacobi orthogonal polynomials (A.20) with the parameters α and β, where

$$J_0^{(\alpha,\beta)}(Z) = 1, \quad J_1^{(\alpha,\beta)}(Z) = \frac{1}{2}[\alpha - \beta + (\alpha + \beta + 2)Z], \quad \dots.$$

The Legendre polynomial chaos becomes a special case of the Jacobi polynomial chaos with $\alpha = \beta = 0$.

In table 5.1, some of the well-known correspondences between the probability distribution of Z and its gPC basis polynomials are listed.

5.1.1 Strong Approximation

The orthogonality (5.2) ensures that the polynomials can be used as basis functions to approximate functions in terms of the random variable Z.

Definition 5.4 (Strong gPC approximation). *Let $f(Z)$ be a function of a random variable Z whose probability distribution is $F_Z(z) = P(Z \le z)$ and support is I_Z. A generalized polynomial chaos approximation in a strong sense is $f_N(Z) \in \mathbb{P}_N(Z)$, where $\mathbb{P}_N(Z)$ is the space of polynomials of Z of degree up to $N \ge 0$, such that $\| f(Z) - f_N(Z) \| \to 0$ as $N \to \infty$, in a proper norm defined on I_Z.*

Table 5.1 Correspondence between the Type of Generalized Polynomial Chaos and Their Underlying Random Variables[a]

	Distribution of Z	gPC basis polynomials	Support
Continuous	Gaussian	Hermite	$(-\infty, \infty)$
	Gamma	Laguerre	$[0, \infty)$
	Beta	Jacobi	$[a, b]$
	Uniform	Legendre	$[a, b]$
Discrete	Poisson	Charlier	$\{0, 1, 2, \dots\}$
	Binomial	Krawtchouk	$\{0, 1, \dots, N\}$
	Negative binomial	Meixner	$\{0, 1, 2, \dots\}$
	Hypergeometric	Hahn	$\{0, 1, \dots, N\}$

[a] $N \geq 0$ is a finite integer.

One obvious strong approximation is the orthogonal projection. Let

$$L^2_{dF_Z}(I_Z) = \{f : I_Z \to \mathbb{R} \mid \mathbb{E}[f^2] < \infty\} \tag{5.6}$$

be the space of all *mean-square integrable* functions with norm $\|f\|_{L^2_{dF_Z}} = (\mathbb{E}[f^2])^{1/2}$. Then, for any function $f \in L^2_{dF_Z}(I_Z)$, we define its Nth-degree *gPC orthogonal projection* as

$$P_N f = \sum_{k=0}^{N} \hat{f}_k \Phi_k(Z), \quad \hat{f}_k = \frac{1}{\gamma_k} \mathbb{E}[f(Z)\Phi_k(Z)]. \tag{5.7}$$

The existence and convergence of the projection follow directly from the classical approximation theory; i.e.,

$$\|f - P_N f\|_{L^2_{dF_Z}} \to 0, \qquad N \to \infty, \tag{5.8}$$

which is also often referred to as *mean-square convergence*. Let $\mathbb{P}_N(Z)$ be the linear space of all polynomials of Z of degree up to N; then the following optimality holds:

$$\|f - P_N f\|_{L^2_{dF_Z}} = \inf_{g \in \mathbb{P}_N(Z)} \|f - g\|_{L^2_{dF_Z}}. \tag{5.9}$$

Though the requirement for convergence (L^2-integrable) is rather mild, the rate of convergence will depend on the smoothness of the function f in terms of Z. The smoother f is, the faster the convergence. These results follow immediately from the classical results reviewed in chapter 3.

When a gPC expansion $f_N(Z)$ of a function $f(Z)$ converges to $f(Z)$ in a strong norm, such as the mean-square norm of (5.8), it implies that $f_N(Z)$ converges to $f(Z)$ in probability, i.e., $f_N \xrightarrow{P} f$, which further implies the convergence in distribution, i.e., $f_N \xrightarrow{d} f$, as $N \to \infty$. (See the discussion of the modes of convergence in section 2.6.)

Example 5.5 (Lognormal random variable). Let $Y = e^X$, where $X \sim \mathcal{N}(\mu, \sigma^2)$ is a Gaussian random variable. The distribution of Y is a *lognormal distribution* whose support is on the nonnegative axis and is widely used in practice to model random variables not allowed to have negative values. Its probability density function is

$$\rho_Y(y) = \frac{1}{y\sigma\sqrt{2\pi}} e^{-\frac{(\ln y - \mu)^2}{2\sigma^2}}. \tag{5.10}$$

To obtain the gPC projection of Y, let $Z \sim \mathcal{N}(0, 1)$ be the standard Gaussian random variable. Then $X = \mu + \sigma Z$ and $Y = f(Z) = e^\mu e^{\sigma Z}$. The Hermite polynomials should be used because of the Gaussian distribution of Z. By following (5.7), we obtain

$$Y_N(Z) = e^{\mu + (\sigma^2/2)} \sum_{k=0}^{N} \frac{\sigma^k}{k!} H_k(Z). \tag{5.11}$$

5.1.2 Weak Approximation

When approximating a function $f(Z)$ with a gPC expansion that converges strongly, e.g., in a mean-square sense, it is necessary to have knowledge of f, that is, the explicit form of f in terms of Z. In practice, however, sometimes only the probability distribution of f is known. In this case, a gPC expansion in terms of Z that converges strongly cannot be constructed because of the lack of information about the dependence of f on Z. However, the approximation can still be made to converge in a weak sense, e.g., in probability. To be precise, the problem can be stated as follows.

Definition 5.6 (Weak gPC approximation). *Let Y be a random variable with distribution function $F_Y(y) = P(Y \leq y)$ and let Z be a (standard) random variable in a set of gPC basis functions. A weak gPC approximation is $Y_N \in \mathbb{P}_N(Z)$, where $\mathbb{P}_N(Z)$ is the linear space of polynomials of Z of degree up to $N \geq 0$, such that Y_N converges to Y in a weak sense, e.g., in probability.*

Obviously, a strong gPC approximation in definition 5.4 implies a weak approximation, not vice versa. We first illustrate the weak approximation via a trivial example and demonstrate that the *gPC weak approximation is not unique*. Let $Y \sim \mathcal{N}(\mu, \sigma^2)$ be a random variable with normal distribution. Naturally we choose $Z \in \mathcal{N}(0, 1)$, a standard Gaussian random variable, and the corresponding Hermite polynomials as the gPC basis. Then a first-order gPC Hermite expansion

$$Y_1(Z) = \mu H_0 + \sigma H_1(Z) = \mu + \sigma Z \tag{5.12}$$

will have precisely the distribution $\mathcal{N}(\mu, \sigma^2)$. Therefore, $Y_1(Z)$ can approximate the distribution of Y *exactly*. However, if all that is known is the distribution of Y, then one cannot reproduce pathwise realizations of Y via $Y_1(Z)$. In fact, $\widetilde{Y}_1(Z) = \mu H_0 - \sigma H_1(Z)$ has the same $\mathcal{N}(\mu, \sigma^2)$ distribution but entirely different pathwise realizations from Y_1.

When Y is an arbitrary random variable with only its probability distribution known, a direct gPC projection in the form of (5.7) is not possible. More specifically, if one seeks an Nth-degree gPC expansion in the form of

$$Y_N = \sum_{k=0}^{N} a_k \Phi_k(Z),$$ (5.13)

with

$$a_k = \mathbb{E}[Y\Phi_k(Z)]/\gamma_k, \qquad 0 \leq k \leq N,$$ (5.14)

where $\gamma_k = \mathbb{E}[\Phi_k^2]$ are the normalization factors, then the expectation in the coefficient evaluation is not properly defined and cannot be carried out, as the dependence between Y and Z is not known. This was first discussed in [120] where a strategy to circumvent the difficulty by using the distribution function $F_Y(y)$ was proposed. The resulting gPC expansion turns out to be the *weak approximation* that we defined here.

By definition, $F_Y : I_Y \to [0, 1]$, where I_Y is the support of Y. Similarly, $F_Z(z) = P(Z \leq z) : I_Z \to [0, 1]$. Since F_Y and F_Z map Y and Z, respectively, to a uniform distribution in $[0, 1]$, we rewrite the expectation in (5.14) in terms of a uniformly distributed random variable in $[0, 1]$. Let $U = F_Y(Y) = F_Z(Z) \sim \mathcal{U}(0, 1)$; then $Y = F_Y^{-1}(U)$ and $Z = F_Z^{-1}(U)$. (The definition of F^{-1} is (2.7).) Now (5.14) can be rewritten as

$$a_k = \frac{1}{\gamma_k} \mathbb{E}_U[F_Y^{-1}(U)\Phi_k(F_Z^{-1}(U))] = \frac{1}{\gamma_k} \int_0^1 F_Y^{-1}(u)\Phi_k(F_Z^{-1}(u))du.$$ (5.15)

This is a properly defined finite integral in $[0, 1]$ and can be evaluated via traditional methods (e.g., Gauss quadrature). Here we use the subscript U in $\mathbb{E}_U$ to make clear that the expectation is over the random variable U.

Alternatively, one can choose to transform the expectation in (5.14) into the expectation in terms of Z by utilizing the fact that $Y = F_Y^{-1}(F_Z(Z))$. Then the expectation in (5.14) can be rewritten as

$$a_k = \frac{1}{\gamma_k} \mathbb{E}_Z[F_Y^{-1}(F_Z(Z))\Phi_k(Z)] = \frac{1}{\gamma_k} \int_{I_Z} F_Y^{-1}(F_Z(z))\Phi_k(z)d F_Z(z).$$ (5.16)

Though (5.15) and (5.16) take different forms, they are mathematically equivalent. The weak convergence of Y_N is established in the following result.

Theorem 5.7. *Let Y be a random variable with distribution $F_Y(y) = P(Y \leq y)$ and $\mathbb{E}(Y^2) < \infty$. Let Z be a random variable with distribution $F_Z(z) = P(Z \leq z)$ and let $\mathbb{E}(|Z|^{2m}) < \infty$ for all $m \in \mathcal{N}$ such that its generalized polynomial chaos basis functions exist with $\mathbb{E}_Z[\Phi_m(Z)\Phi_n(Z)] = \delta_{mn}\gamma_n, \forall m, n \in \mathcal{N}$. Let*

$$Y_N = \sum_{k=0}^{N} a_k \Phi_k(Z),$$ (5.17)

where

$$a_k = \frac{1}{\gamma_k} \mathbb{E}_Z[F_Y^{-1}(F_Z(Z))\Phi_k(Z)], \qquad 0 \leq k \leq N.$$ (5.18)

Then Y_N converges to Y in probability; i.e.,

$$Y_N \xrightarrow{P} Y, \qquad N \to \infty. \tag{5.19}$$

Also, $Y_N \xrightarrow{d} Y$ in distribution.

Proof. Let

$$\widetilde{Y} \triangleq G(Z) = F_Y^{-1}(F_Z(Z)),$$

where $G \triangleq F_Y^{-1} \circ F_Z : I_Z \to I_Y$. Obviously, $\widetilde{Y}$ has the same probability distribution as that of Y, i.e., $F_{\widetilde{Y}} = F_Y$, and we have $\widetilde{Y} \overset{P}{=} Y$ and $\mathbb{E}[\widetilde{Y}^2] < \infty$. This immediately implies

$$\mathbb{E}[\widetilde{Y}^2] = \int_{I_Y} y^2 d F_Y(y)$$

$$= \int_0^1 \left(F_Y^{-1}(u) \right)^2 du$$

$$= \int_{I_Z} \left(F_Y^{-1}(F_Z(z)) \right)^2 d F_Z(z) < \infty.$$

Therefore, $\widetilde{Y} = G(Z) \in L^2_{dF_Z}(I_Z)$. Since (5.17) and (5.18) is the orthogonal projection of $\widetilde{Y}$ by the Nth-degree gPC basis, the strong convergence of Y_N to $\widetilde{Y}$ in mean square implies convergence in probability, i.e., $Y_N \xrightarrow{P} \widetilde{Y}$ as $N \to \infty$. Since $\widetilde{Y} \overset{P}{=} Y$, the main conclusion follows. Since convergence in probability implies convergence in distribution, $Y_N \xrightarrow{d} Y$. The completes the proof.

Example 5.8 (Approximating beta distribution by gPC Hermite expansion).
Let the probability distribution of Y be a beta distribution with probability density function $\rho(y) \propto (1 - y)^{\alpha}(1 + y)^{\beta}$. In this case, if one chooses the corresponding Jacobi polynomials as the gPC basis function, then the first-order gPC expansion can satisfy the distribution exactly. However, suppose one chooses to employ the gPC Hermite expansion in terms of $Z \sim \mathcal{N}(0, 1)$; then a weak approximation can still be obtained via the procedure discussed here. All is needed is a numerical approximation of the integral (5.15) or (5.16). In figure 5.1, the convergence in PDF is shown for different orders of the gPC Hermite expansion. Numerical oscillations near the corners of the distributions can be clearly seen, resembling Gibbs oscillations. Note that the support of Y is in $[-1, 1]$ and is quite different from the support of Z (which is $\mathbb{R}$).

Example 5.9 (Approximating exponential distribution by gPC Hermite expansion). Now let us assume that the distribution of Y is an exponential distribution with the PDF $\rho(y) \propto e^{-y}$. Figure 5.2 shows the convergence of PDF by the gPC Hermite expansions. Note that the first-order expansion, which results in a Gaussian distribution, is entirely different from the target exponential distribution. As the order of expansion is increased, the approximation improves. In this case, if one chooses the corresponding gPC basis, i.e., the Laguerre polynomials, then the first-order expansion can produce the exponential distribution *exactly*.

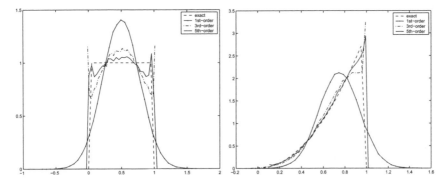

Figure 5.1 Approximating beta distributions by gPC Hermite expansions: convergence of probability density functions with increasing order of expansions. Left: approximation of uniform distribution $\alpha = \beta = 0$. Right: approximation of beta distribution with $\alpha = 2$, $\beta = 0$. (More details are in [120].)

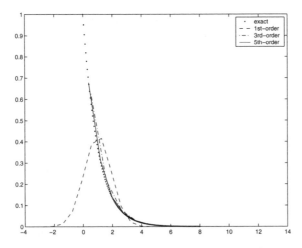

Figure 5.2 Approximating an exponential distribution by gPC Hermite expansions: convergence of probability density function with increasing order of expansions.

Example 5.10 (Approximating Gaussian distribution by gPC Jacobi expansion). Let us assume that the distribution of Y is the standard Gaussian $\mathcal{N}(0, 1)$ and use the gPC Jacobi expansion to approximate the distribution. The convergence in PDF is shown in figure 5.3, where both the Legendre polynomials and the Jacobi polynomials with $\alpha = \beta = 10$ are used. We observe some numerical oscillations when using the gPC Legendre expansion. Again, if we use the corresponding gPC basis for Gaussian distribution, the Hermite polynomials, then the first-order expansion $Y_1 = H_1(Z) = Z$ will have precisely the desired $\mathcal{N}(0, 1)$ distribution.

It is also worth noting that the approximations by gPC Jacobi chaos with $\alpha = \beta = 10$ are quite good, even at the first order. This implies that the beta distribution with $\alpha = \beta = 10$ is very close to the Gaussian distribution $\mathcal{N}(0, 1)$. However,

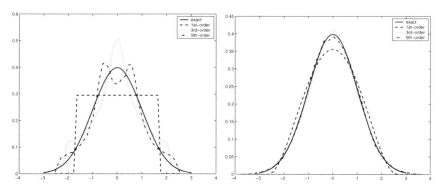

Figure 5.3 Approximations of Gaussian distribution by gPC Jacobi expansions: convergence of probability density functions with increasing order of expansions. Left: approximation by gPC Jacobi polynomials with $\alpha = \beta = 0$ (Legendre polynomials). Right: approximation by gPC Jacobi polynomials with $\alpha = \beta = 10$.

a distinct feature of the beta distribution is that it is strictly bounded in a close interval. This suggests that in practice when one needs a distribution that is close to Gaussian but with strict bounds, mostly because of concerns from a physical or mathematical point of view, then the beta distribution can be a good candidate. More details on this approximation can be found in appendix B.

From these examples it is clear that when the corresponding gPC polynomials for a given distribution function can be constructed, particularly for the well-known cases listed in table 5.1, it is best to use these basis polynomials because a proper first-order expansion can produce the given distribution exactly. Using other types of polynomials can still result in a convergent series at the cost of inducing approximation errors and more complex gPC representation.

5.2 DEFINITION IN MULTIPLE RANDOM VARIABLES

When more than one independent random variables are involved, multivariate gPC expansion is required. Let $Z = (Z_1, \ldots, Z_d)$ be a random vector with mutually independent components and distribution $F_Z(z_1, \ldots, z_d) = P(Z_1 \leq z_1, \ldots, Z_d \leq z_d)$. For each $i = 1, \ldots, d$, let $F_{Z_i}(z_i) = P(Z_i \leq z_i)$ be the marginal distribution of Z_i, whose support is I_{Z_i}. Mutual independence among all Z_i implies that $F_Z(z) = \prod_{i=1}^{d} F_{Z_i}(z_i)$ and $I_Z = I_{Z_1} \times \cdots \times I_{Z_d}$. Also, let $\{\phi_k(Z_i)\}_{k=0}^{N} \in \mathbb{P}_N(Z_i)$ be the univariate gPC basis functions in Z_i of degree up to N. That is,

$$\mathbb{E}\left[\phi_m(Z_i)\phi_n(Z_i)\right] = \int \phi_m(z)\phi_n(z)dF_{Z_i}(z) = \delta_{mn}\gamma_m, \quad 0 \leq m, n \leq N. \quad (5.20)$$

Let $\mathbf{i} = (i_1, \ldots, i_d) \in \mathbb{N}_0^d$ be a multi-index with $|\mathbf{i}| = i_1 + \cdots + i_d$. Then, the d-variate Nth-degree gPC basis functions are the products of the univariate gPC polynomials (5.20) of total degree less than or equal to N; i.e.,

$$\Phi_{\mathbf{i}}(Z) = \phi_{i_1}(Z_1) \cdots \phi_{i_d}(Z_d), \quad 0 \leq |\mathbf{i}| \leq N. \quad (5.21)$$

It follows immediately from (5.20) that

$$\mathbb{E}\left[\Phi_{\mathbf{i}}(Z)\Phi_{\mathbf{j}}(Z)\right] = \int \Phi_{\mathbf{i}}(z)\Phi_{\mathbf{j}}(z)d\,F_Z(z) = \gamma_{\mathbf{i}}\delta_{\mathbf{ij}}, \qquad (5.22)$$

where $\gamma_{\mathbf{i}} = \mathbb{E}[\Phi_{\mathbf{i}}^2] = \gamma_{i_1}\cdots\gamma_{i_d}$ are the normalization factors and $\delta_{\mathbf{ij}} = \delta_{i_1 j_1}\cdots\delta_{i_d j_d}$ is the d-variate Kronecker delta function. It is obvious that the span of the polynomials is $\mathbb{P}_N^d$, the linear space of all polynomials of degree at most N in d variables,

$$\mathbb{P}_N^d(Z) = \left\{ p: I_Z \to \mathbb{R} \,\middle|\, p(Z) = \sum_{|\mathbf{i}|\leq N} c_{\mathbf{i}}\Phi_{\mathbf{i}}(Z) \right\}, \qquad (5.23)$$

whose dimension is

$$\dim \mathbb{P}_N^d = \binom{N+d}{N}. \qquad (5.24)$$

The space of *homogeneous gPC*, following Wiener's notion of *homogeneous chaos*, is the space spanned by the gPC polynomials in (5.21) of degree precisely N; that is,

$$\mathcal{P}_N^d(Z) = \left\{ p: I_Z \to \mathbb{R} \,\middle|\, p(Z) = \sum_{|\mathbf{i}|=N} c_{\mathbf{i}}\Phi_{\mathbf{i}}(Z) \right\} \qquad (5.25)$$

and

$$\dim \mathcal{P}_N^d = \binom{N+d-1}{N}. \qquad (5.26)$$

The d-variate gPC projection follows the univariate projection in a direct manner. Let $L_{d\,F_Z}^2(I_Z)$ be the space of all mean-square integrable functions of Z with respect to the measure $d\,F_Z$; that is,

$$L_{d\,F_Z}^2(I_Z) = \left\{ f: I_Z \to \mathbb{R} \,\middle|\, \mathbb{E}[f^2(Z)] = \int_{I_Z} f^2(z)d\,F_Z(z) < \infty \right\}. \qquad (5.27)$$

Then for $f \in L_{d\,F_Z}^2(I_Z)$, its Nth-degree *gPC orthogonal projection* is defined as

$$P_N f = \sum_{|\mathbf{i}|\leq N} \hat{f}_{\mathbf{i}}\Phi_{\mathbf{i}}(Z), \qquad (5.28)$$

where

$$\hat{f}_{\mathbf{i}} = \frac{1}{\gamma_{\mathbf{i}}}\mathbb{E}[f\Phi_{\mathbf{i}}] = \frac{1}{\gamma_{\mathbf{i}}}\int f(z)\Phi_{\mathbf{i}}(z)d\,F_Z(z), \qquad \forall|\mathbf{i}| \leq N. \qquad (5.29)$$

The classical approximation theory can be readily applied to obtain

$$\|f - P_N f\|_{L_{d\,F_Z}^2} \to 0, \qquad N \to \infty, \qquad (5.30)$$

and

$$\|f - P_N f\|_{L_{d\,F_Z}^2} = \inf_{g\in\mathbb{P}_N^d} \|f - g\|_{L_{d\,F_Z}^2}. \qquad (5.31)$$

Table 5.2 An example of graded lexicographic ordering of the multi-index $\mathbf{i}$ in $d = 4$ dimensions

$|\mathbf{i}|$	Multi-index $\mathbf{i}$	Single index k
0	$(0\,0\,0\,0)$	1
1	$(1\,0\,0\,0)$	2
	$(0\,1\,0\,0)$	3
	$(0\,0\,1\,0)$	4
	$(0\,0\,0\,1)$	5
2	$(2\,0\,0\,0)$	6
	$(1\,1\,0\,0)$	7
	$(1\,0\,1\,0)$	8
	$(1\,0\,0\,1)$	9
	$(0\,2\,0\,0)$	10
	$(0\,1\,1\,0)$	11
	$(0\,1\,0\,1)$	12
	$(0\,0\,2\,0)$	13
	$(0\,0\,1\,1)$	14
	$(0\,0\,0\,2)$	15
3	$(3\,0\,0\,0)$	16
	$(2\,1\,0\,0)$	17
	$(2\,0\,1\,0)$	18
	$\cdots$	$\cdots$

Up to this point, the mutual independence among the components of Z has not been used explicitly, in the sense that all the expositions are made on $dF_Z(z)$ and I_Z in a general manner and do not require the properties of $I_Z = I_{Z_1} \times \cdots \times I_{Z_d}$ and $dF_Z(z) = dF_{Z_1}(z_1) \cdots dF_{Z_d}(z_d)$, which are direct consequences of independence. This implies that the above presentation of gPC is applicable to more general cases.

Although clear for the formulation, the multi-index notations employed here are cumbersome to manipulate in practice. It is therefore preferable to use a single index to express the gPC expansion. A popular choice is the *graded lexicographic order*, where $\mathbf{i} > \mathbf{j}$ if and only if $|\mathbf{i}| \geq |\mathbf{j}|$ and the first nonzero entry in the difference, $\mathbf{i} - \mathbf{j}$, is positive. Though other choices exist, the graded lexicographic order is the most widely adopted one in practice. The multi-index can now be ordered in an ascending order following a single index. For example, for a $(d = 4)$-dimensional case, the graded lexicographic order is shown in table 5.2.

Let us also remark that the polynomial space (5.23) is not the only choice. Another option is to keep the highest polynomial order for up to N in *each direction*. That is,

$$\widetilde{\mathbb{P}}_N^d(Z) = \left\{ p : I_Z \to \mathbb{R} \,\middle|\, p(Z) = \sum_{|\mathbf{i}|_0 \leq N} c_\mathbf{i} \Phi_\mathbf{i}(Z) \right\}, \tag{5.32}$$

where $|\mathbf{i}|_0 = \max_{1 \leq j \leq d} i_j$. This kind of space is friendly to theoretical analysis (e.g., [8]), as properties of one dimension can be more easily extended. On the other hand,

$\dim \widetilde{\mathbb{P}}_N^d = N^d$. And for large dimensions, the number of basis functions grows too fast. Therefore, this space is usually not adopted in practical computations.

5.3 STATISTICS

When a sufficiently accurate gPC expansion for a given function $f(Z)$ is available, one has in fact an *analytical* representation of f in terms of Z. Therefore, practically all statistical information can be retrieved from the gPC expansion in a straightforward manner, either analytically or with minimal computational effort.

Let us use a random process to illustrate the idea. Consider a process $f(t, Z)$, $Z \in \mathbb{R}^d$ and $t \in T$, where T is an index set. For any fixed $t \in T$, let

$$f_N(t, Z) = \sum_{|\mathbf{i}| \leq N} \hat{f}_{\mathbf{i}}(t) \Phi_{\mathbf{i}}(Z) \in \mathbb{P}_N^d$$

be an Nth-degree gPC approximation of $f(t, Z)$; i.e., $f_N \approx f$ in a proper sense (e.g., mean square) for any $t \in T$. Then the *mean* of f can be approximated as

$$\mu_f(t) \triangleq \mathbb{E}[f(t, Z)] \approx \mathbb{E}[f_N(t, Z)] = \int \left(\sum_{|\mathbf{i}| \leq N} \hat{f}_{\mathbf{i}}(t) \Phi_{\mathbf{i}}(z) \right) dF_Z(z) = \hat{f}_{\mathbf{0}}(t),$$

(5.33)

following the orthogonality of the gPC basis functions (5.22). The *second moments*, e.g., the *covariance function*, can be approximated by, for any $t_1, t_2 \in T$,

$$\begin{aligned} C_f(t_1, t_2) &\triangleq \mathbb{E}[(f(t_1, Z) - \mu_f(t_1))(f(t_2, Z) - \mu_f(t_2))] \\ &\approx \mathbb{E}[(f_N(t_1, Z) - \hat{f}_{\mathbf{0}}(t_1))(f_N(t_2, Z) - \hat{f}_{\mathbf{0}}(t_2))] \\ &= \sum_{0 < |\mathbf{i}| \leq N} [\gamma_{\mathbf{i}} \hat{f}_{\mathbf{i}}(t_1) \hat{f}_{\mathbf{i}}(t_2)]. \end{aligned}$$

(5.34)

The *variance* of f can be obviously approximated by, for any $t \in T$,

$$\text{var}(f(t, Z)) = \mathbb{E}\left[\left(f(t, Z) - \mu_f(t) \right)^2 \right] \approx \sum_{0 < |\mathbf{i}| \leq N} \left[\gamma_{\mathbf{i}} \hat{f}_{\mathbf{i}}^2(t) \right].$$

(5.35)

Other statistical quantities of f can also be readily approximated by applying their definitions directly to the gPC approximation f_N.

Chapter Six

Stochastic Galerkin Method

In this chapter we discuss the generalized polynomial chaos (gPC) Galerkin method for solving stochastic systems. We first introduce the main idea via a general stochastic partial differential equation (PDE) and then illustrate more detailed properties of the method by applying it to several representative problems.

6.1 GENERAL PROCEDURE

Again, without loss of generality, we utilize the stochastic PDE system (4.13). For a physical domain $D \subset \mathbb{R}^\ell$, $\ell = 1, 2, 3$, and $T > 0$, consider

$$
\begin{cases}
u_t(x, t, Z) = \mathcal{L}(u), & D \times (0, T] \times \mathbb{R}^d, \\
\mathcal{B}(u) = 0, & \partial D \times [0, T] \times \mathbb{R}^d, \\
u = u_0, & D \times \{t = 0\} \times \mathbb{R}^d,
\end{cases}
\tag{6.1}
$$

where again $\mathcal{L}$ is the differential operator, $\mathcal{B}$ is the boundary condition operator, u_0 is the initial condition, and $Z = (Z_1, \dots, Z_d) \in \mathbb{R}^d$, $d \geq 1$, are a set of mutually independent random variables characterizing the random inputs to the governing equation. For ease of presentation, let us consider a scalar equation where

$$
u(x, t, Z) : \bar{D} \times [0, T] \times \mathbb{R}^d \to \mathbb{R}.
$$

For a system of equations, the gPC expansion will be applied to each component of u individually.

Let $\{\Phi_\mathbf{k}(Z)\}$ be the gPC basis functions satisfying

$$
\mathbb{E}[\Phi_\mathbf{i}(Z)\Phi_\mathbf{j}(Z)] = \delta_{\mathbf{ij}}\gamma_\mathbf{i}
\tag{6.2}
$$

and let $\mathbb{P}_N^d(Z)$ be the space of all polynomials of $Z \in \mathbb{R}^d$ of degree up to N. Then the gPC projection of the solution is, for any fixed (x, t),

$$
u_N(x, t, Z) = \sum_{|\mathbf{i}|=0}^{N} \hat{u}_\mathbf{i}(x, t)\Phi_\mathbf{i}(Z), \quad \hat{u}_\mathbf{i}(x, t) = \frac{1}{\gamma_\mathbf{i}}\mathbb{E}[u(x, t, Z)\Phi_\mathbf{i}(Z)].
\tag{6.3}
$$

Though this is the optimal (in the $L^2_{dF_Z}$ sense) approximation in $\mathbb{P}_N^d$, it is not of practical use since the projection requires knowledge of the unknown solution.

The stochastic Galerkin procedure is a straightforward extension of the classical Galerkin approach for deterministic equations. That is, we seek a solution in $\mathbb{P}_N^d$ such that the residue of (6.1) is orthogonal to the space $\mathbb{P}_N^d$. By utilizing the gPC orthogonal basis functions (6.2), we obtain the following procedure: for any x and t, we seek $v_N \in \mathbb{P}_N^d$ in the form of

$$v_N(x, t, Z) = \sum_{|\mathbf{i}|=0}^{N} \hat{v}_{\mathbf{i}}(x, t) \Phi_{\mathbf{i}}(Z), \qquad (6.4)$$

such that for all $\mathbf{k}$ satisfying $|\mathbf{k}| \leq N$,

$$\begin{cases} \mathbb{E}[\partial_t v_N(x, t, Z) \Phi_{\mathbf{k}}(Z)] = \mathbb{E}[\mathcal{L}(v_N) \Phi_{\mathbf{k}}], & D \times (0, T], \\ \mathbb{E}[\mathcal{B}(v_N) \Phi_{\mathbf{k}}] = 0, & \partial D \times [0, T], \\ \hat{v}_{\mathbf{k}} = \hat{u}_{0,\mathbf{k}}, & D \times \{t = 0\}, \end{cases} \qquad (6.5)$$

where $\hat{u}_{0,\mathbf{k}} = \mathbb{E}[u_0 \Phi_{\mathbf{k}}]/\gamma_{\mathbf{k}}$ are the gPC projection coefficients for the initial condition. Upon evaluating the expectations in (6.5), the dependence in Z disappears. The result is a system of (usually coupled) deterministic equations. The size of the system is $\dim \mathbb{P}_N^d = \binom{N+d}{N}$.

6.2 ORDINARY DIFFERENTIAL EQUATIONS

Let us use the ordinary differential equation in example 4.4 to illustrate the main steps of the gPC Galerkin method.

$$\frac{du}{dt}(t, Z) = -\alpha(Z)u, \qquad u(t = 0, Z) = \beta,$$

where the initial condition is assumed to be deterministic (for simplicity). We also assume that the random rate constant follows a normal distribution; i.e., $\alpha \sim \mathcal{N}(\mu, \sigma^2)$. The corresponding gPC basis will be the Hermite polynomials. Since α is the only random input, we need only univariate gPC Hermite expansion $\{H_k(Z)\}_{k=0}^{N}$, $N > 0$, where $Z \sim \mathcal{N}(0, 1)$ is the standard normal random variable with zero mean and unit variance. The constant α can be expressed as $\alpha = \mu + \sigma Z$. Or, in a more general form,

$$\alpha_N(Z) = \sum_{i=0}^{N} a_i H_i(Z),$$

where

$$a_0 = \mu, \qquad a_1 = \sigma, \qquad a_i = 0, \quad i > 1.$$

Usually α_N is an approximation of α. However, in this case it is an exact expression as long as $N \geq 1$. Similarly, the initial condition has a trivial gPC projection,

$$\beta_N = \sum_{i=0}^{N} b_i H_i(Z),$$

where

$$b_0 = \beta, \qquad b_i = 0, \quad i > 0,$$

which is exact for $N > 0$.

Let

$$v_N(t, Z) = \sum_{i=0}^{N} \hat{v}_i(t)\Phi_i(Z)$$

be the Nth-degree gPC approximation we seek. The gPC Galerkin procedure results in

$$\mathbb{E}\left[\frac{dv_N}{dt}H_k\right] = \mathbb{E}[-\alpha_N v_N H_k], \qquad \forall k = 0, \ldots, N.$$

Upon substituting in the gPC expression for α_N and v_N, we obtain

$$\frac{d\hat{v}_k}{dt} = -\frac{1}{\gamma_k}\sum_{i=0}^{N}\sum_{j=0}^{N} a_i \hat{v}_j e_{ijk} \qquad \forall k = 0, \ldots, N, \tag{6.6}$$

where

$$e_{ijk} = \mathbb{E}[H_i(Z)H_j(Z)H_k(Z)], \qquad 0 \leq i, j, k \leq N, \tag{6.7}$$

are constants. Like the normalization factors γ_k, these constants can be evaluated prior to any computations. In fact, for Hermite polynomials these constants can be evaluated analytically,

$$\gamma_k = k! \qquad k \geq 0, \tag{6.8}$$

$$e_{ijk} = \frac{i!j!k!}{(s-i)!(s-j)!(s-k)!}, \qquad s \geq i, j, k, \text{ and } 2s = i + j + k \text{ is even.} \tag{6.9}$$

For other types of gPC basis functions, the analytical expressions for the constants may not exist. In such cases, one can use numerical quadrature rules with a sufficient number of points to compute the constants numerically *but exactly* since the integrands are of polynomial form.

System (6.6) is thus a system of deterministic ordinary differential equations for the coefficients $\{\hat{v}_k(t)\}$ with initial conditions

$$\hat{v}_k(0) = b_k, \qquad 0 \leq k \leq N. \tag{6.10}$$

The size of the system is $N + 1$, and the equations are coupled. Classical numerical methods, e.g., Runge-Kutta methods, can be applied, and usually the coupling of the system does not pose serious numerical challenges. (More details on numerical studies of this problem can be found in [120].)

We can also rewrite the system in a compact form by using vector notation. By taking the summation over i in (6.6) and defining

$$A_{jk} = -\frac{1}{\gamma_k}\sum_{i=0}^{N} a_i e_{ijk},$$

we let $\mathbf{A} = (A_{jk})_{0 \leq j,k \leq N}$ be a $(N + 1) \times (N + 1)$ matrix. Denote $\mathbf{v}(t) = (\hat{v}_0, \ldots, \hat{v}_N)^T$; then (6.6) can be written as

$$\frac{d\mathbf{v}}{dt}(t) = \mathbf{A}^T \mathbf{v}, \qquad \mathbf{v}(0) = \mathbf{b}, \tag{6.11}$$

where $\mathbf{b} = (b_0, \ldots, b_N)^T$.

6.3 HYPERBOLIC EQUATIONS

Let us now consider a simple linear wave equation

$$\frac{\partial u(x,t,Z)}{\partial t} = c(Z)\frac{\partial u(x,t,Z)}{\partial x}, \qquad x \in (-1,1), \quad t > 0, \tag{6.12}$$

where $c(Z)$ is a random transport velocity that is a function of a random variable $Z \in \mathbb{R}$. For now we will leave the distribution of Z unspecified and study the general properties of the resulting gPC Galerkin system. The initial condition is given by

$$u(x,0,Z) = u_0(x,Z). \tag{6.13}$$

The boundary conditions are more complicated, as they depend on the sign of the random transport velocity $c(Z)$. A well-posed set of boundary conditions is given by

$$\begin{aligned} u(1,t,Z) &= u_R(t,Z), & c(Z) > 0, \\ u(-1,t,Z) &= u_L(t,Z), & c(Z) < 0. \end{aligned} \tag{6.14}$$

The interesting issue to understand is how to properly pose the boundary conditions for the gPC Galerkin system.

Again an univariate gPC expansion is sufficient. For ease of analysis, let us use the normalized gPC basis functions,

$$\mathbb{E}[\Phi_i(Z)\Phi_j(Z)] = \delta_{ij}, \qquad 0 \le i,j \le N.$$

Note that the normalization only requires dividing the nonnormalized basis by the square root of the normalization constants. It facilitates the theoretical analysis only. In practical implementations, one does not need to normalize the basis. With the gPC Galerkin method, we seek, for any (x,t),

$$v_N(x,t,Z) = \sum_{i=0}^{N} \hat{v}_i(x,t)\Phi_i(Z) \tag{6.15}$$

and conduct the projection

$$\mathbb{E}\left[\frac{\partial v_N(x,t,Z)}{\partial t}\Phi_k(Z)\right] = \mathbb{E}\left[c(Z)\frac{\partial v_N(x,t,Z)}{\partial x}\Phi_k(Z)\right]$$

for each of the first $N+1$ gPC basis $k = 0, \ldots, N$. We obtain

$$\frac{\partial \hat{v}_k(x,t)}{\partial t} = \sum_{i=0}^{N} a_{ik}\frac{\partial \hat{v}_i(x,t)}{\partial x}, \qquad k = 0, \ldots, N, \tag{6.16}$$

where

$$a_{ik} = \mathbb{E}[c(Z)\Phi_i(Z)\Phi_k(Z)], \qquad 0 \le i,k \le N. \tag{6.17}$$

This is now a coupled system of wave equations of size $N+1$, where the coupling is through the random wave speed. If we denote by $\mathbf{A}$ the $(N+1) \times (N+1)$

matrix whose entries are $\{a_{ik}\}_{0 \leq i,k \leq N}$, then by definition $a_{ik} = a_{ki}$ and $\mathbf{A} = \mathbf{A}^T$ is symmetric. Let $\mathbf{v} = (\hat{v}_0, \ldots, \hat{v}_N)^T$ be a vector of length $N + 1$; then system (6.16) can be written as

$$\frac{\partial \mathbf{v}(x, t)}{\partial t} = \mathbf{A} \frac{\partial \mathbf{v}(x, t)}{\partial x}. \tag{6.18}$$

It is now clear that system (6.18) is *symmetric hyperbolic*. Therefore, a complete set of real eigenvalues and eigenfunctions exists. Moreover, we can understand the signs of the eigenvalues, which indicate the direction of the wave for the gPC Galerkin system (6.18), based on the signs of wave direction in the original system (6.12).

Theorem 6.1. *Consider the gPC Galerkin system* (6.18) *derived from the original system* (6.12). *Then if $c(Z) \geq 0$ (respectively, $c(Z) \leq 0$) for all Z, then the eigenvalues of $\mathbf{A}$ are all nonnegative (respectively, nonpositive); if $c(Z)$ changes sign, i.e., $c(Z) > 0$ for some Z and $c(Z) < 0$ for some other Z, then $\mathbf{A}$ has both positive and negative eigenvalues for sufficiently high gPC expansion order N.*

The proof can be found in [48].

A less trivial issue is how to impose the inflow-outflow boundary conditions for the hyperbolic system (6.18), especially when the wave speed changes signs in the original system (6.12). Note that the explicit information about the sign of the wave speed disappears in the gPC Galerkin system (6.18). Because (6.18) is symmetric hyperbolic, we can diagonalize the system and then impose boundary conditions based on the sign of the eigenvalues.

Since $\mathbf{A}$ is symmetric, there exists an orthogonal matrix $\mathbf{S}^T = \mathbf{S}^{-1}$ such that $\mathbf{S}^T \mathbf{A} \mathbf{S} = \Lambda$, where Λ is a diagonal matrix whose entries are the eigenvalues of $\mathbf{A}$; i.e.,

$$\Lambda = \mathrm{diag}(\lambda_0, \ldots, \lambda_{j_+}, \ldots, \lambda_{j_-}, \ldots, \lambda_N).$$

Here the positive eigenvalues occupy indices $j = 0, \ldots, j_+$, the negatives ones occupy indices $j = j_-, \ldots, N$, and the rest, if they exist, are zeros. Obviously, $j_+, j_- \leq N$.

Denote $\mathbf{q} = (q_0, \ldots, q_N)^T = \mathbf{S}^T \mathbf{v}$, i.e., $q_j(x, t) = \sum_{k=0}^{N} s_{kj} \hat{v}_k(x, t)$, where s_{jk} are the entries for $\mathbf{S}$; then we obtain

$$\frac{\partial \mathbf{q}(x, t)}{\partial t} = \Lambda \frac{\partial \mathbf{q}(x, t)}{\partial x}. \tag{6.19}$$

The boundary conditions of this diagonal system are determined by the sign of the eigenvalues; i.e., we need to specify

$$q_j(1, t) = \sum_{k=0}^{N} s_{kj} \hat{u}_k(1, t), \qquad j = 0, \ldots, j_+,$$

$$q_j(-1, t) = \sum_{k=0}^{N} s_{kj} \hat{u}_k(-1, t), \qquad j = j_-, \ldots, N.$$

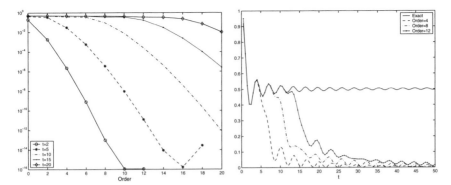

Figure 6.1 Convergence property of the gPC Galerkin solution to the wave problem (6.12). Left: error convergence with respect to the order of gPC expansion at different times. Right: evolution of the solution in mean-square norm in time at different gPC orders.

Here the coefficients $\hat{u}_k$ at the boundaries are determined by the exact gPC projection of the boundary conditions of u, i.e., u_R and u_L. Subsequently, the boundary conditions for the gPC Galerkin system of equations (6.16) are specified as

$$\mathbf{v}(1, t) = \mathbf{Sq}(1, t), \quad \mathbf{v}(-1, t) = \mathbf{Sq}(-1, t).$$

For vanishing eigenvalues, if they exist, no boundary conditions are required.

It can be shown that the solution of the gPC Galerkin system (6.18) converges to the exact solution. In fact, the following error bound was established (see theorem 2.2 in [48]),

$$\mathbb{E}\left[\|u - v_N\|_2^2\right] \leq \frac{C}{N^{2m-1}} t, \tag{6.20}$$

where $\|\cdot\|$ is the standard L^2 norm in the physical domain $(-1, 1)$, C is a constant independent of N, t is time, and $m > 0$ is a real constant depending on the smoothness of u in terms of Z.

A notable feature of the error bound is that the error depends on time in a linear manner. This implies that for any fixed gPC expansion order N, the error will grow linearly in time. This can be seen in figure 6.1, where the gPC Galerkin solution is applied to a simpler version of (6.12) with $c(Z) = Z$, a uniform random variable in $(-1, 1)$, *periodic boundary conditions in space* in a domain $0 \leq x \leq 2\pi$, and initial condition $u(x, 0, Z) = \cos(x)$. The exact solution is $u_{\text{ex}} = \cos(x - Zt)$. On the left side of figure 6.1, while we again see the exponential error convergence as the gPC order is increased, the error is bigger at a larger time t and requires higher expansion orders to reach the converging regime. The time dependence becomes more evident on the right side of figure 6.1, where the evolution of the mean-square norm of the solutions are plotted. We observe that the gPC Galerkin solutions deviate from the exact solution after a certain time. The time at which the accuracy is lost, i.e., errors become $O(1)$, is roughly proportional to the order of the gPC expansion.

It is important to realize the following facts.

- The linear growth of error is not a result of the boundary condition treatment. This is obvious from the results in figure 6.1 because they are obtained from a problem with a periodic condition (and require no special treatment of the boundary conditions).
- Furthermore, the linear growth of error is *not a result of the Galerkin approach*. In fact, a direct gPC orthogonal projection of the exact solution $u_{\text{ex}} = \cos(x - Zt)$ would require more and more basis function in order to keep a fixed projection error. This is because as far as the expansion in Z is concerned, the time t behaves as a wave number. A larger time t thus requires finer representations. This is the fundamental property of approximation theory.

6.4 DIFFUSION EQUATIONS

Let us consider a time-dependent stochastic diffusion equation

$$
\begin{cases}
\dfrac{\partial u}{\partial t}(x, t, Z) = \nabla_x \cdot (\kappa(x, Z)\nabla_x u) + f(x, t), & x \in D, \ t \in (0, T], \\
u(x, 0, Z) = u_0(x), \quad u|_{\partial D} = 0.
\end{cases}
\tag{6.21}
$$

Here we use ∇_x to explicitly specify that the differentiation is in the physical coordinates x. We assume that the only source of uncertainty is from the diffusivity field κ, which causes coupling of the resulting gPC Galerkin system of equations. Uncertainty in the source term f and initial condition u_0 will not cause coupling and can be dealt with easily. We assume the diffusivity field takes a form

$$
\kappa(x, Z) = \hat{\kappa}_0(x) + \sum_{i=1}^{d} \hat{\kappa}_i(x)Z_i,
\tag{6.22}
$$

where $\hat{\kappa}_i(x)$ are deterministic functions obtained by applying a parameterization procedure (e.g., the Karhunen-Loeve expansion) to the diffusivity field and that $Z = (Z_1, \ldots, Z_d)$ are mutually independent random variables with specified probability distributions. Alternatively, we can write (6.22) as

$$
\kappa(x, Z) = \sum_{i=0}^{d} \hat{\kappa}_i(x)Z_i,
\tag{6.23}
$$

where $Z_0 \equiv 1$ is fixed. For the problem to be well posed, we require

$$
\kappa(x, Z) \geq \kappa_{\min} > 0, \qquad \forall x, Z.
\tag{6.24}
$$

Such a requirement obviously excludes probability distributions of Z that take negative values with nonzero probability, e.g., Gaussian distributions.

Again we seek an Nth-degree gPC approximation

$$
v_N(t, x, Z) = \sum_{|\mathbf{k}|=0}^{N} \hat{v}_{\mathbf{k}}(t, x)\Phi_{\mathbf{k}}(Z),
$$

where $\mathbb{E}[\Phi_i(Z)\Phi_j(Z)] = \delta_{ij}$. Here we again choose to normalize the gPC basis first. For ease of exposition, let us adopt the single-index notation discussed in section 5.2 and rewrite the gPC expansion as

$$v_N(t, x, Z) = \sum_{i=1}^{M} \hat{v}_i(t, x)\Phi_i(Z), \quad M = \binom{N+d}{N}, \tag{6.25}$$

where the index i is determined by a proper ordering, e.g., the graded lexicographic ordering, of the multi-index $\mathbf{i} \in \mathbb{N}_0^d$.

Upon substituting (6.23) and the gPC expansion (6.25) into the governing equation (6.21) and projecting the resulting equation onto the subspace spanned by the first M gPC basis polynomials, we obtain, for all $k = 1, \dots, M$,

$$\frac{\partial \hat{v}_k}{\partial t}(t, x) = \sum_{i=0}^{d} \sum_{j=1}^{M} \nabla_x \cdot (\hat{\kappa}_i(x)\nabla_x \hat{v}_j) e_{ijk} + \hat{f}_k(t, x)$$

$$= \sum_{j=1}^{M} \nabla_x \cdot (a_{jk}(x)\nabla_x \hat{v}_j) + \hat{f}_k(t, x), \tag{6.26}$$

where

$$e_{ijk} = \mathbb{E}[Z_i \Phi_j \Phi_k] = \int z_i \Phi_j(z)\Phi_k(z) dF_Z(z), \quad 0 \le i \le d, \ 1 \le j, k \le M,$$

$$a_{jk}(x) = \sum_{i=0}^{d} \hat{\kappa}_i(x) e_{ijk}, \quad 1 \le j, k \le M, \tag{6.27}$$

and $\hat{f}_k(t, x)$ are the gPC projection coefficients for the source term $f(x, t)$. (For the simple case of deterministic f considered here, $\hat{f}_1 = f$ and $\hat{f}_k = 0$ for $k > 1$.)

The gPC Galerkin system (6.26) is a coupled system of diffusion equations. It can be put into a compact form by using vector matrix notation. Let us denote $\mathbf{v} = (\hat{v}_1, \dots, \hat{v}_M)^T$, $\mathbf{f} = (\hat{f}_1, \dots, \hat{f}_M)^T$, and $\mathbf{A}(x) = (a_{jk})_{1 \le j, k \le M}$. By definition, $\mathbf{A} = \mathbf{A}^T$ is symmetric. The system (6.26) can be written as

$$\frac{\partial \mathbf{v}}{\partial t}(t, x) = \nabla_x \cdot [\mathbf{A}(x)\nabla_x \mathbf{v}] + \mathbf{f}, \quad (t, x) \in (0, T] \times D,$$

$$\mathbf{v}(0, x) = \mathbf{v}_0(x), \quad \mathbf{v}|_{\partial D} = 0, \tag{6.28}$$

where $\mathbf{v}_0(x) = (\hat{u}_{0,1}, \dots, \hat{u}_{0,M})^T$ is the gPC projection coefficient vector of the initial condition $u_0(x)$ in (6.21). For the deterministic initial condition considered here, $\hat{u}_{0,1} = u_0(x)$ and $\hat{u}_{0,k} = 0$ for $k > 1$.

The coupling of the diffusion terms in (6.28) does not pose a serious problem if the system is solved explicitly in time. However, an explicit time integration usually imposes a severe restriction on the size of the time step because of concerns about numerical stability. To circumvent this difficulty, one can employ a semi-implicit scheme where the diagonal terms of $\mathbf{A}$ are treated implicitly and the off-diagonal terms of $\mathbf{A}$ are treated explicitly. This results in a naturally *uncoupled* system to solve with *no loss of accuracy in time integration*. For example, a first-oder Euler forward-backward semi-implicit scheme takes the form

$$\frac{\mathbf{v}^{n+1} - \mathbf{v}^n}{\Delta t} - \nabla_x \cdot \left[\mathbf{D}(x) \nabla_x \mathbf{v}^{n+1} \right] = \nabla_x \cdot \left[\mathbf{S}(x) \nabla_x \mathbf{v}^n \right] + \mathbf{f}^{n+1}, \qquad (6.29)$$

where the superscript n denotes numerical solutions at time level t_n, Δt is the time step, and

$$\mathbf{D} = \text{diag}(\mathbf{A}), \quad \mathbf{A} = \mathbf{D} + \mathbf{S}.$$

Similarly, if we consider the steady-state counterpart of (6.21),

$$-\nabla_x \cdot (\kappa(x, Z) \nabla_x u(x, Z)) = f(x), \quad x \in D; \qquad u(x, Z)|_{\partial D} = 0, \qquad (6.30)$$

we find that the gPC Galerkin system is

$$-\nabla_x \cdot [\mathbf{A}(x) \nabla_x \mathbf{v}] = \mathbf{f}, \quad x \in D; \qquad \mathbf{v}|_{\partial D} = 0. \qquad (6.31)$$

This is a coupled system of elliptic equations. By using the separation of diagonal and off-diagonal terms of $\mathbf{A}$, an efficient iterative scheme can be designed to solve the system as an uncoupled system of equations. These algorithms were first proposed in [119, 122] and later analyzed in [128].

6.5 NONLINEAR PROBLEMS

The above examples all involve linear problems. This does not imply that the gPC Galerkin method can be applied only to linear problems. (In fact, as far as the random space is concerned, none of the examples are linear because the randomness in the equations is all in a multiplicative manner.)

Let us consider the Burgers' equation from the supersensitivity example in section 1.1.1 to illustrate application of the gPC Galerkin method to nonlinear problems.

$$\begin{cases} u_t + u u_x = \nu u_{xx}, & x \in [-1, 1], \\ u(-1) = 1 + \delta(Z), & u(1) = -1, \end{cases} \qquad (6.32)$$

where $\delta(Z) > 0$ is a random perturbation to the left boundary condition at $(x = -1)$ and $\nu > 0$ is the viscosity. Again this requires a one-dimensional gPC expansion. We seek

$$v_N(x, t, Z) = \sum_{i=0}^{N} \hat{v}_i(x, t) \Phi_i(Z)$$

such that

$$\mathbb{E}\left[\frac{\partial v_N}{\partial t} \Phi_k \right] + \mathbb{E}\left[v_N \frac{\partial v_N}{\partial x} \Phi_k \right] = \nu \mathbb{E}\left[\frac{\partial^2 v_N}{\partial x^2} \Phi_k \right], \quad k = 0, \ldots, N.$$

By substituting v_N into the equation and using the orthogonality relation of the basis functions, we obtain

$$\frac{\partial \hat{v}_k}{\partial t} + \frac{1}{\gamma_k} \sum_{i=0}^{N} \sum_{j=0}^{N} \hat{v}_i \frac{\partial \hat{v}_j}{\partial x} e_{ijk} = \nu \frac{\partial^2 \hat{v}_k}{\partial x^2}, \qquad k = 0, \ldots, N, \qquad (6.33)$$

where $e_{ijk} = \mathbb{E}[\Phi_i \Phi_j \Phi_k]$ are constants and $\gamma_k = \mathbb{E}[\Phi_k^2]$ are the normalization constants (which will be 1 if the basis functions are normalized).

This is a coupled system of equations where each equation resembles the original Burgers' equation and the coupling is through the nonlinear term. The classical semi-implicit scheme can be applied to solve the system in time, where the nonlinear coupling terms are treated explicitly and the diffusion terms implicitly. For more details, see [123].

The nonlinear term uu_x in the Burgers' equation is in quadratic form and results in a gPC projection

$$\mathbb{E}\left[v_N \frac{\partial v_N}{\partial x} \Phi_k\right] = \sum_{i=0}^{N} \sum_{j=0}^{N} \hat{v}_i \frac{\partial \hat{v}_j}{\partial x} \mathbb{E}[\Phi_i \Phi_j \Phi_k]$$

that can be easily evaluated as long as the term is treated as being explicit in time. In many cases, however, nonlinear terms in a system do not take polynomial form and a direct gPC projection is not straightforward. For example, let us consider the projection of a nonlinear term e^u, where u is the unknown solution. A gPC Galerkin projection requires us to evaluate

$$\mathbb{E}\left[e^{v_N} \Phi_k\right] = \int e^{\sum_i \hat{v}_i \Phi_i(z)} \Phi_k(z) dF_Z(z), \qquad (6.34)$$

where v_N is the Nth-degree gPC approximation of u. It is clear that the integral over z cannot be separated from the summation over i, as in the case of polynomial-type nonlinearity.

A feasible treatment for such kinds of nonlinearity is to approximate the integral (6.34) numerically. To this end, one can employ a quadrature rule, or a cubature rule in multivariate cases, with sufficient accuracy. That is,

$$\mathbb{E}\left[e^{v_N} \Phi_k\right] \approx \sum_{j=1}^{Q} e^{v_N(z^{(j)})} \Phi_k(z^{(j)}) w^{(j)}, \qquad (6.35)$$

where $z^{(j)}$ and $w^{(j)}$ are the nodes and weights of the integration rule in the domain defined by the integral. Note that since $v_N(Z)$ takes a known polynomial form, the evaluation of e^{v_N} at any node is a simple exercise in polynomial evaluation.

Chapter Seven

Stochastic Collocation Method

In this chapter we discuss the basic ideas behind the stochastic collocation (SC) method, also referred to as the probabilistic collocation method (PCM). The collocation methods are a popular choice for complex systems where well-established deterministic codes exist. We first clarify the notion of stochastic collocation, for the purposes of this book, and then discuss the major numerical approaches. As in the rest of the book, we discuss only the fundamental aspects of SC and leave the more advanced research issues untouched. This is particularly true for this chapter because SC has undergone rapid development after its systematic introduction in [118].

7.1 DEFINITION AND GENERAL PROCEDURE

In deterministic numerical analysis, collocation methods are those that require the residue of the governing equations to be zero at discreet nodes in the computational domain. The nodes are called *collocation* points. The same definition can be extended to stochastic simulations. Let us use the stochastic partial differential equation (PDE) system (4.13) again to explain the idea,

$$\begin{cases} u_t(x, t, Z) = \mathcal{L}(u), & D \times (0, T] \times I_Z, \\ \mathcal{B}(u) = 0, & \partial D \times [0, T] \times I_Z, \\ u = u_0, & D \times \{t = 0\} \times I_Z, \end{cases} \tag{7.1}$$

where $I_Z \subset \mathbb{R}^d$, $d \geq 1$, is the support of Z. For any given x and t, let $w(\cdot, Z)$ be a numerical approximation of u. In general, $w(\cdot, Z) \approx u(\cdot, Z)$ in a proper sense in I_Z, and the system (7.1) cannot be satisfied for all Z after substituting u for w.

Let $\Theta_M = \{Z^{(j)}\}_{j=1}^M \subset I_Z$ be a set of (prescribed) nodes in the random space, where $M \geq 1$ is the number of nodes. Then in the collocation method, for all $j = 1, \ldots, M$, we enforce (7.1) at the node $Z^{(j)}$ by solving

$$\begin{cases} u_t(x, t, Z^{(j)}) = \mathcal{L}(u), & D \times (0, T], \\ \mathcal{B}(u) = 0, & \partial D \times [0, T], \\ u = u_0, & D \times \{t = 0\}. \end{cases} \tag{7.2}$$

It is easy to see that for each j, (7.2) is a deterministic problem because the value of the random parameter Z is fixed. Therefore, solving the system poses no difficulty provided one has a well-established deterministic algorithm. Let $u^{(j)} = u(\cdot, Z^{(j)})$, $j = 1, \ldots, M$, be the solution of the above problem. The result of solving (7.2) is an ensemble of deterministic solutions $\{u^{(j)}\}_{j=1}^{M}$. And one can apply various post-processing operations to the ensemble to extract useful information about $u(Z)$.

From this point of view, all classical sampling methods belong to the class of collocation methods. For example, in *Monte Carlo sampling*, the nodal set Θ_M is generated randomly according to the distribution of Z, and the ensemble average is used to estimate the solution statistics, e.g., mean and variance. In *deterministic sampling methods*, the nodal set is typically the nodes of a cubature rule (i.e., quadrature rule in multidimensional space) defined on I_Z such that one can use the integration rule defined by the cubature to estimate the solution statistics. Convergence of these classical sampling methods is then based on the convergence of solution statistics, e.g., moments, resulting in convergence in a weak measure such as convergence in distribution.

In this book we do not label the classical sampling methods as stochastic collocation. Instead we reserve the term "stochastic collocation" for the type of collocation methods that result in a strong convergence, e.g., mean-square convergence, to the true solution. This is typically achieved by utilizing the classical multivariate approximation theory to strategically locate the collocation nodes to construct a polynomial approximation to the solution.

Definition 7.1 (Stochastic collocation). *Let $\Theta_M = \{Z^{(j)}\}_{j=1}^{M} \subset I_Z$ be a set of (prescribed) nodes in the random space, where $M \geq 1$ is the number of nodes, and let $\{u^{(j)}\}_{j=1}^{M}$ be the solution of the governing equation (7.2). Then find $w(Z) \in \Pi(Z)$ in a proper polynomial space $\Pi(Z)$ such that $w(Z)$ is an approximation to the true solution $u(Z)$ in the sense that $\|w(Z) - u(Z)\|$ is sufficiently small in a strong norm defined on I_Z.*

Convergence of stochastic collocation thus requires

$$\|w(Z) - u(Z)\| \to 0, \qquad M \to \infty,$$

where the norm is to be determined and is typically an L^p norm.

As of the writing of this book, there exist two major approaches for high-order stochastic collocation: the *interpolation approach* and the *discrete projection approach* (the *pseudospectral approach*).

7.2 INTERPOLATION APPROACH

Interpolation is a natural approach to the stochastic collocation problem defined in definition 7.1. The problem can now be posed as follows: given the nodal set $\Theta_M \subset I_Z$ and $\{u^{(j)}\}_{j=1}^{M}$, find a polynomial $w(Z) \in \Pi(Z)$ such that $w(Z^{(j)}) = u^{(j)}$ for all $j = 1, \ldots, M$.

The goal can be easily accomplished, at least in principle. One way is to use a *Lagrange interpolation approach*. That is, let

$$w(Z) = \sum_{j=1}^{M} u(Z^{(j)}) L_j(Z), \qquad (7.3)$$

where

$$L_j(Z^{(i)}) = \delta_{ij}, \qquad 1 \le i, j \le M, \tag{7.4}$$

are the Lagrange interpolating polynomials. The approach, albeit straightforward in formulation, can become nontrivial in practice. This is mostly due to the fact that unlike the situation in univariate interpolation, where ample mathematical theory exists, many fundamental issues of multivariate Lagrange interpolation (when $d > 1$) are not clear. Issues such as the existence of Lagrange interpolating polynomials for any given set of notes are not well understood.

The other way is a *matrix inversion approach*, where we prescribe the polynomial interpolating basis first. For example, let us use a set of gPC polynomial bases $\Phi_{\mathbf{k}}(Z)$ and construct

$$w_N(Z) = \sum_{|\mathbf{k}|=0}^{N} \hat{w}_{\mathbf{k}} \Phi_{\mathbf{k}}(Z)$$

as the gPC approximation of $u(Z)$. The interpolation condition $w(Z^{(j)}) = u^{(j)}$ results in the following problem for the unknown expansion coefficients

$$\mathbf{A}^T \hat{\mathbf{w}} = \mathbf{u},$$

where

$$\mathbf{A} = (\Phi_{\mathbf{k}}(Z^{(j)})), \qquad 0 \le |\mathbf{k}| \le N, \ \ 1 \le j \le M,$$

is the Vandermonde-like coefficient matrix, $\hat{\mathbf{w}}$ is the vector of the expansion coefficients, and $\mathbf{u} = (u(Z^{(1)}), \dots, u(Z^{(M)}))^T$. To prevent the problem from becoming underdetermined, we require the number of collocation points not to be smaller than the number of gPC expansion terms, i.e., $M \ge \binom{N+d}{N}$. The advantage of the matrix inversion approach is that the interpolating polynomials are prescribed and well defined. Once the nodal set is given, the existence of the interpolation can always be determined in the spirit of determining whether the determinant of $\mathbf{A}$ is zero. However, an important and very practical concern is the accuracy of the interpolation. Even though the interpolation has no error at the nodal points, error can become wild between the nodes. This is particularly true in high-dimensional spaces. Here again, we find rigorous analysis lacking and no general (and sound) guideline for choosing the location of the nodes. Many ad hoc choices do exist, for example, those based on design of experiments (DoE) principles. However, none has become satisfactory for general purposes.

Since univariate interpolation is a well-studied topic, one solution to multivariate interpolation is to employ a univariate interpolation and then fill up the entire space dimension by dimension. By doing so the properties and error estimates of univariate interpolation can be retained as much as possible. In fact, the aforementioned two approaches, the Lagrange interpolation and matrix inversion approaches, are direct *conceptual extensions* of the univariate interpolation techniques in section 3.4. Let us recall that in the univariate case $d = 1$, i.e., $Z \in \mathbb{R}$. Let $(Z^{(1)}, \dots, Z^{(N+1)})$ be a set of distinct nodes and let $(u^{(1)}, \dots, u^{(N+1)})$ be the solution at the nodes. Then an interpolation polynomial $\Pi_N f(Z)$ that interpolates a given function $f(Z)$ can be constructed either in the Lagrange form (3.42) or by inverting the Vandermonde matrix to obtain (3.43). Note that the two are equivalent because of the uniqueness of univariate interpolation. It is also understood that the interpolating nodes offering high accuracy are the zeros of the orthogonal polynomials $\{\Phi_k(Z)\}$.

7.2.1 Tensor Product Collocation

For multivariate cases with $d > 1$, for any $1 \leq i \leq d$, let Q_{m_i} be an interpolating operator such that

$$Q_{m_i}[f] = \Pi_{m_i} f(Z_i) \in \mathbb{P}_{m_i}(Z_i)$$

is an interpolating polynomial of degree m_i, for a given function f in the Z_i variable by using $m_i + 1$ distinct nodes in the set $\Theta_1^{m_i} = \{Z_i^{(1)}, \ldots, Z_i^{(m_i)}\}$. Then the most straightforward approach to interpolating f in the entire space $I_Z \subset \mathbb{R}^d$ is to use a tensor product approach. That is,

$$Q_M = Q_{m_1} \otimes \cdots \otimes Q_{m_d}, \tag{7.5}$$

and the nodal set is

$$\Theta_M = \Theta_1^{m_1} \times \cdots \times \Theta_1^{m_d}, \tag{7.6}$$

where the total number of nodes is $M = m_1 \times \cdots \times m_d$.

By using the tensor product construction, all the properties of the underlying one-dimensional interpolation scheme can be retained. And error estimate in the entire space can be easily derived. For example, let us assume that the number of points in each dimension is a constant; i.e., $m_1 = \cdots = m_d = m$, and that the one-dimensional interpolation error in each dimension $1 \leq i \leq d$ follows

$$(I - Q_{m_i})[f] \propto m^{-\alpha},$$

where the constant $\alpha > 0$ depends on the smoothness of the function f. Then the overall interpolation error also follows the same convergence rate

$$(I - Q_M)[f] \propto m^{-\alpha}.$$

However, if we measure the convergence in terms of the total number of points, $M = m^d$ in this case, then

$$(I - Q_M)[f] \propto M^{-\alpha/d}, \qquad d \geq 1.$$

For large dimensions $d \gg 1$, the rate of convergence deteriorates drastically and we observe very slow convergence, if there is any, in terms of the total number of collocation points. Moreover, the total number of points,

$$M = m^d,$$

grows very fast for large d. This poses a numerical challenge because each collocation point requires a simulation of the full-scale underlying deterministic system, which can be time-consuming. This is the well-known *curse of dimensionality*. For this reason, tensor product construction is mostly used for low-dimensional problems with d typically less than 5. A detailed theoretical analysis for stochastic diffusion equations can be found in [7].

7.2.2 Sparse Grid Collocation

An alternative approach is Smolyak sparse grids. A detailed presentation of the Smolyak construction, originally proposed in [96], is beyond the scope of this entry-level textbook, and we refer interested readers to the many more recent studies. It is sufficient, for the purposes of this book, to know that the Smolyak sparse grids are still based on tensor product construction but are only a subset of the full tensor grids. The construction takes the following form ([114]):

$$Q_N = \sum_{N-d+1 \leq |\mathbf{i}| \leq N} (-1)^{N-|\mathbf{i}|} \cdot \binom{d-1}{N-|\mathbf{i}|} \cdot \left(Q_{i_1} \otimes \cdots \otimes Q_{i_d} \right), \qquad (7.7)$$

where $N \geq d$ is an integer denoting the *level* of the construction. Though the expression is rather complex, (7.7) is nevertheless a combination of the subsets of the full tensor grids. The nodal set, the *sparse grids*, is

$$\Theta_M = \bigcup_{N-d+1 \leq |\mathbf{i}| \leq N} (\Theta_1^{i_1} \times \cdots \times \Theta_1^{i_d}). \qquad (7.8)$$

Again it is clear that this is the union of a collection of subsets of the full tensor grids. Unfortunately, there is usually no explicit formula to determine the total number of nodes M in terms of d and N.

Because the construction in (7.7) employs one-dimensional interpolations with various numbers of nodes, it is preferable that one-dimensional nodal sets be *nested*. That is, the one-dimensional nodal sets satisfy

$$\Theta_1^i \subset \Theta_1^j, \qquad i < j. \qquad (7.9)$$

If this condition is met, then the total number of nodes in (7.8) can reach a minimum. However, in practice, since one-dimensional nodes are typically the zeros of orthogonal polynomials, the nested condition (7.9) is usually not satisfied.

One popular choice of nested grids is Clenshaw-Curtis nodes, which are the extrema of Chebyshev polynomials and are defined as, for any $1 \leq i \leq d$,

$$Z_i^{(j)} = -\cos\frac{\pi(j-1)}{m_i^k - 1}, \qquad j = 1, \ldots, m_i^k, \qquad (7.10)$$

where an additional index is introduced via the superscript k. With a slight abuse of notation, we will use k to index the point sets instead of using the total number of points m_i. Let the point sets double with an increasing index of $k > 1$, i.e., $m_i^k = 2^{k-1} + 1$, and define $m_i^1 = 1$ and $Z_i^{(1)} = 0$. It is easy to see that because of the doubling of the nodes we have $\Theta_1^k \subset \Theta_1^{k+1}$ and that the sets are nested. The additional index k here is often referred to as the *level* of Clenshaw-Curtis grids. The higher the level, the finer the grids. For a more detailed discussion of Clenshaw-Curtis nodes, see [27].

By using Clenshaw-Curtis grids as one-dimensional nodes, the Smolyak construction (7.7) can be expressed in terms of the *level* k as well. Let $N = d + k$, where $k \geq 0$, and then the "nestedness" of the base one-dimensional nodes can be retained, $\Theta_k \subset \Theta_{k+1}$. Again here we do not use the total number of nodes M, which does not have an explicit and closed form expression in terms of d and k,

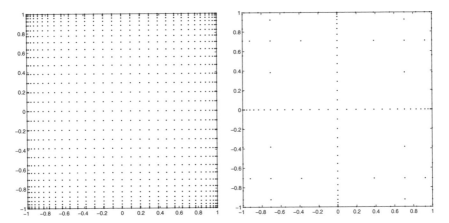

Figure 7.1   Two-dimensional ($d = 2$) nodes based on the same one-dimensional extrema of
Chebyshev polynomials at level $k = 5$. Left: tensor grids. The total number of
points is 1089. Right: Smolyak sparse grids. The total number of nodes is 145.

to index the sets. In the more interesting case of high-dimensional spaces, the total
number of points satisfies the following estimate:

$$M = \#\Theta_k \sim 2^k d^k / k!, \qquad d \gg 1. \tag{7.11}$$

It has been shown ([9]) that interpolation through the Clenshaw-Curtis sparse grid
interpolation is exact if the function is in $\mathbb{P}_k^d$. (In fact, the polynomial space for
which the interpolation is exact is slightly bigger than $\mathbb{P}_k^d$.) For large dimensions
$d \gg 1$, $\dim \mathbb{P}_k^d = \binom{d+k}{d} \sim d^k / k!$. Therefore, the number of points from (7.11) is
about 2^k more and the factor is *independent* of the dimension d. For this reason, the
Clenshaw-Curtis–based sparse grids construction is sometimes regarded as optimal
in high dimensions. There have been extensive studies on the approximation properties of sparse grids, particularly those based on Clenshaw-Curtis nodes. Here we
cite one of the early results from [9]. For functions in space $F_d^\ell = \{ f : [-1, 1]^d \to$
$\mathbb{R} | \partial^{|i|} f$ continuous, $i_j \le \ell, \forall j \}$, the interpolation error follows

$$\| I - Q_M \|_\infty \le C_{d,\ell} M^{-\ell} (\log M)^{(\ell+2)(d-1)+1},$$

where M is the total number of nodes. Compared to that for tensor grids, the curse
of dimensionality, albeit still present, is lessened.

    An example of two-dimensional sparse grids is shown in figure 7.1, where we
observe a significant reduction in the number of nodes.

7.3 DISCRETE PROJECTION: PSEUDOSPECTRAL APPROACH

Another approach to achieving the goal of stochastic collocation, as defined in
definition 7.1, is to conduct *discrete projection*, or the *pseudospectral approach*
(as it is termed in [116]). To this end, let us first recall the notion of the quadrature

rule in section 3.5 and extend it to multivariate space. Often termed the *cubature rule*, it is an *integration rule* that seeks to approximate an integral

$$\int f(z)\, dF_Z(z), \qquad z \in \mathbb{R}^d, \quad d > 1, \tag{7.12}$$

by

$$U^Q[f] \triangleq \sum_{j=1}^{Q} f(z^{(j)})\alpha^{(j)}, \qquad Q \geq 1, \tag{7.13}$$

where $(z^{(j)}, \alpha^{(j)})$, $j = 1, \ldots, Q$, are the nodes and their corresponding weights. The integration rule is *convergent* if it converges to the integral (7.12) as $Q \to \infty$. Typically, the accuracy of an integration is measured by *polynomial exactness*. An integration rule of degree m implies that the approximation (7.13) is exact for any integrand f that is a polynomial of degree up to m and is not exact for at least one polynomial of degree $m + 1$. Hereafter we will freely interchange the notation of the integration rule and the cubature rule, with an understanding that in univariate cases it is reduced to the quadrature rule.

To conduct discrete gPC projection, we recall the exact orthogonal gPC projection of $u(Z)$, the solution of (7.1),

$$u_N(Z) = P_N u = \sum_{|\mathbf{k}|=0}^{N} \hat{u}_{\mathbf{k}} \Phi_{\mathbf{k}}(Z), \tag{7.14}$$

where the expansion coefficients are obtained as

$$\hat{u}_{\mathbf{k}} = \frac{1}{\gamma_{\mathbf{k}}} \mathbb{E}[u(Z)\Phi_{\mathbf{k}}(Z)] = \frac{1}{\gamma_{\mathbf{k}}} \int u(z)\Phi_{\mathbf{k}}(z)\, dF_Z(z), \qquad \forall |\mathbf{k}| \leq N, \tag{7.15}$$

where $\gamma_{\mathbf{k}} = \mathbb{E}[\Phi_{\mathbf{k}}^2]$ are the normalization constants of the basis.

The idea of discrete projection is to approximate the integrals in the expansion coefficients (7.15) of the continuous generalized polynomial chaos (gPC) projection (7.14) by an integration rule. The discrete projection of the solution of (7.1) is

$$w_N(Z) = \sum_{|\mathbf{k}|=0}^{N} \hat{w}_{\mathbf{k}} \Phi_{\mathbf{k}}(Z), \tag{7.16}$$

where the expansion coefficients are

$$\hat{w}_{\mathbf{k}} = \frac{1}{\gamma_{\mathbf{k}}} U^Q[u(Z)\Phi_{\mathbf{k}}(Z)] = \frac{1}{\gamma_{\mathbf{k}}} \sum_{j=1}^{Q} u(z^{(j)})\Phi_{\mathbf{k}}(z^{(j)})\alpha^{(j)}. \tag{7.17}$$

It is clear that by using the cubature rule U^Q the coefficients $\{\hat{w}_{\mathbf{k}}\}$ are approximations to the exact projection coefficients $\{\hat{u}_{\mathbf{k}}\}$ in (7.15). Subsequently, the discrete projection $w_N(Z)$ approximates the continuous projection $u_N(Z)$ of (7.14). Moreover, if the cubature rule is convergent, then $\hat{w}_{\mathbf{k}}$ converges to $\hat{u}_{\mathbf{k}}$ as $Q \to \infty$, and w_N and u_N become identical. The following result is then straightforward.

Proposition 7.2. *For $u(Z) \in L^2_{dF_Z}(I_Z)$, let $u_N(Z)$ be the gPC projection defined in (7.14) and (7.15) and let $w_N(Z)$ be the discrete gPC projection defined in (7.16) and (7.17). Assume that the cubature rule U^Q used in (7.17) is convergent; then as $Q \to \infty$, $\hat{w}_{\mathbf{k}} \to \hat{u}_{\mathbf{k}}$, for all $|\mathbf{k}| \leq N$, and*

$$w_N(Z) \to u_N(Z), \quad \forall Z. \tag{7.18}$$

The error induced by w_N can be easily separated by the triangular inequality.

Proposition 7.3. *For $u(Z) \in L^2_{dF_Z}(I_Z)$, let $u_N(Z)$ be the gPC projection defined in (7.14) and (7.15) and let $w_N(Z)$ be the discrete gPC projection defined in (7.16) and (7.17). Then,*

$$\|w_N(Z) - u(Z)\|_{L^2_{dF_Z}} \leq \|u_N(Z) - u(Z)\|_{L^2_{dF_Z}} + \|w_N(Z) - u_N(Z)\|_{L^2_{dF_Z}}. \tag{7.19}$$

The first term of the error is the gPC projection error induced by using finite-order (Nth-order) polynomials. The second term of the error is the difference between the continuous gPC projection and the discrete projection and is caused by using a cubature rule with finite accuracy. It can be expressed as

$$\epsilon_N^Q \triangleq \|w_N(Z) - u_N(Z)\|_{L^2_{dF_Z}} = \left(\sum_{|\mathbf{k}|=0}^{N} (w_{\mathbf{k}} - u_{\mathbf{k}})^2 \gamma_{\mathbf{k}} \right)^{1/2}, \tag{7.20}$$

and is termed "aliasing error" in [116], by following the similar nomenclature from classical deterministic spectral methods (cf. [13, 46]).

Similar to the interpolation approach, the construction of gPC expansion via (7.16) and (7.17) can also be considered a postprocessing step after all the computations are finished at the cubature nodes. A distinct feature of the discrete projection approach is that one can compute only the coefficients that are important for a given problem without evaluating the rest of the expansion coefficients. This may happen, for example, when global sensitivity is required for some input random variables Z. This is in contrast to the gPC Galerkin method, where all the gPC coefficients are coupled and solved simultaneously.

Since the main task in the discrete gPC projection approach is to approximate the integrals in (7.15), the problem of multivariate polynomial approximation is transformed to a problem of multivariate integration, where the accuracy of the chosen integration rule is critical. Compared to multivariate interpolation, which is used by the stochastic interpolation collocation approach, there exist, relatively speaking, more results on multivariate integration, which is nevertheless a challenging and very active research topic.

7.3.1 Structured Nodes: Tensor and Sparse Tensor Constructions

Since Gauss quadrature rules offer high accuracy for univariate integrations, it is natural to extend them to multivariate integrations. The most straightforward way of constructing high-order integration rules is to extend quadrature rules (in univariate cases) to high-dimensional spaces by using tensor construction. Let U^{m_i} be a Gauss quadrature rule in the Z_i direction of $Z = (Z_1, \ldots, Z_d)$, $d > 1$, with a nodal set $\Theta_1^{m_i}$ consisting of an $m_i \geq 1$ number of nodes. Let us assume that it is exact for all polynomials in $\mathbb{P}^{2m_i-1}(Z_i)$. Then a tensor construction is

$$U^Q = U^{m_1} \otimes \cdots \otimes U^{m_d}.$$

Obviously, the nodal set is

$$\Theta^Q = \Theta_1^{m_1} \times \cdots \times \Theta_1^{m_d},$$

whose total number of nodes is $Q = m_1 \times \cdots \times m_d$. This integration rule is exact for all polynomials in

$$\mathbb{P}^{2m_1-1}(Z_1) \otimes \cdots \otimes \mathbb{P}^{2m_d-1}(Z_d).$$

Though easy to construct and of high accuracy, the problem is again the rapid growth of the total number of points in high-dimensional random spaces. If we use an equal number of nodes in all directions, $m_1 = \cdots = m_d = m$, then the total number of nodes is $Q = m^d$. For $d \gg 1$, this can be a staggeringly large number. (Again let us keep in mind that at each node the full-scale deterministic problem needs to be solved.) Consequently, the tensor product approach is mostly used at lower dimensions, e.g., $d \leq 5$.

To reduce the total number of nodes while keeping most of the high accuracy offered by Gauss quadrature, the Smolyak sparse grids construction can be employed, similarly to the sparse interpolation discussed in section 7.2.2,

$$U^Q = \sum_{N-d+1\leq|\mathbf{m}|\leq N} (-1)^{N-|\mathbf{m}|} \cdot \binom{d-1}{N-|\mathbf{m}|} \cdot \left(U^{m_1} \otimes \cdots \otimes U^{m_d}\right), \qquad (7.21)$$

where $N \geq d$ is an integer denoting the *level* of construction. The grid set, the *sparse grids*, is

$$\Theta^Q = \bigcup_{N-d+1\leq|\mathbf{m}|\leq N} (\Theta_1^{m_1} \times \cdots \times \Theta_1^{m_d}). \qquad (7.22)$$

Again it is clear that this is the union of a collection of subsets of the full tensor grids. Usually there is no closed-form explicit formula for the total number of nodes Q.

Depending on the choice of Gauss quadrature in one dimension, there are a variety of sparse grid constructions. And they offer different accuracy. Many of the constructions are based on the Clenshaw-Curtis rule in one dimension, whose properties are closely examined in [108]. Here we will not engage in further indepth discussion of the technical details. Interested readers should see, for example, [9, 39, 83, 84].

7.3.2 Nonstructured Nodes: Cubature

The study of cubature rules is a relatively old topic but is still actively pursued. The goal is to construct a cubature rule with a high degree of polynomial exactness and a lesser number of nodes. To achieve this, structured nodes are usually not considered and many studies are based on geometric considerations. Most of the rules are given in explicit formulas, regarding the node locations and their corresponding weights.

Depending on the required accuracy for the discrete gPC projection, one can choose a proper cubature with an affordable number of nodes for simulations. For extensive reviews and collections of available cubature rules, see, for example, [21, 49, 99].

It is worth pointing out that the classical error estimate, in terms of error bounds and such, is nontrivial to carry out for cubature rules. Consequently, the accuracy of cubature rules is almost always classified by their polynomial exactness.

7.4 DISCUSSION: GALERKIN VERSUS COLLOCATION

While the gPC expansion provides a solid framework for stochastic computations, a natural question to ask is, For a given practical stochastic system, should one use the Galerkin method or the collocation method?

The advantage of stochastic collocation is clear—ease of implementation. The algorithms are straightforward: (1) choose a set of nodes according to either multivariate interpolation theory or integration theory; (2) run deterministic code at each node; and (3) postprocess to construct the gPC polynomials, via either the interpolation approach (section 7.2) or the discrete projection approach (section 7.3). The applicability of stochastic collocation is not affected by the complexity or nonlinearity of the original problem so long as one can develop a reliable deterministic solver. The executions of the deterministic algorithm at each node are completely independent of each other and embarrassingly parallel. For these reasons, stochastic collocation methods have become very popular.

The stochastic Galerkin method, on the other hand, is more cumbersome to implement. The Galerkin system (6.5) needs to be derived, and the resulting equations for the expansion coefficients are almost always coupled. Hence new codes need to be developed to deal with the larger, coupled system of equations. When the original problem (6.1) takes highly complex and nonlinear form, the explicit derivation of the gPC equations can be nontrivial—sometimes impossible.

However, an important issue to keep in mind is the accuracy of the methods. The stochastic Galerkin approach ensures that the residue of the stochastic governing equations is orthogonal to the linear space spanned by the gPC polynomials, as in (6.5). In this sense, the accuracy is optimal. On the other hand, stochastic collocation approaches, with no error at the nodes, introduce errors either because of the interpolation scheme (if interpolation collocation is used) or because of the integration rule (if discrete projection collocation is used). Both errors are caused by introduction of the nodal sets and can be classified as *aliasing errors*. Though in one dimension, aliasing error can be kept at the same order as the error of the finite order Galerkin method, in multidimensional spaces the aliasing errors can be much more significant. Roughly speaking, at a fixed accuracy, which is usually measured in terms of the polynomial exactness of the approximation, all of the existing collocation methods require the solution of a (much) larger number of equations than that required by the gPC Galerkin method, especially for higher-dimensional random spaces. This suggests that the gPC Galerkin method offers the most accurate solutions involving the least number of equations in multidimensional random spaces, even though the equations are coupled.

The exact cost comparison between the Galerkin and the collocation methods depends on many factors including error analysis, which is largely unknown, and even on coding efforts involved in developing a stochastic Galerkin code. However, it is safe to state that for large-scale simulations where a single deterministic computation is time-consuming, the stochastic Galerkin method should be preferred (because of the lesser number of equations) whenever (1) the coupling of gPC Galerkin equations does not incur much additional computational and coding effort. For example, for Navier-Stokes equations with random boundary/initial conditions the evaluations of the coupling terms are trivial ([121]) or (2) efficient solvers can be developed to effectively decouple the gPC Galerkin system. For example, for stochastic diffusion equations decoupling can be achieved in the manner in (6.29).

Another factor that should be taken into account as part of the effort in implementing the stochastic Galerkin method is that the properties of the Galerkin system may not be clear, even when the baseline deterministic system is well understood. And this may affect our design of numerical algorithms for the Galerkin system. A simple example is the linear wave equation with random wave speed in section 6.3.

Chapter Eight

Miscellaneous Topics and Applications

The discussions up to this point have involved numerical algorithms, mostly based on generalized polynomial chaos (gPC), for propagating uncertainty from inputs to outputs. Here we will discuss several related topics regarding variations and applications of gPC methods. More specifically, we will consider efficient algorithms for

- random geometry, where the uncertain input is in the specification of the computational domain,
- parameter estimation, i.e., how to estimate the probability distribution of the input parameters, and
- uncertainty in models and how to "correct" it by using available measurement data.

For the second and third topics, the help of experimental measurement is required. gPC-based stochastic methods can improve the accuracy of existing methods at virtually no simulation cost.

Unlike those in the previous chapters of this book, the topics in this chapter are closer to research issues. However, here we will only briefly discuss these topics with a focus on their *direct connections* with gPC methods. More in-depth general discussions and literature reviews can be found in the references.

8.1 RANDOM DOMAIN PROBLEM

Throughout this book, we have always assumed that the computational domain is fixed and contains no uncertainty. In practice, however, it can be a major source of uncertainty, as in many applications the physical domain cannot be determined precisely. The problem with uncertain geometry, e.g., a rough boundary, has been studied in areas such as wave scattering with many specially designed techniques. (See, for example, a review in [113].) For general-purpose partial differential equations (PDEs), however, numerical techniques in uncertain domains are less developed. Here we discuss general numerical approaches by following one of the earlier systematic studies in [130].

Let $D(\omega)$, $\omega \in \Omega$, be a random domain whose boundary $\partial D(\omega)$ is the source of randomness. Since $\partial D(\omega)$ is a random process, we first seek to parameterize it by a function of a finite number of independent random variables. This parameterization procedure is required for the ensuing stochastic simulations. Its implementation, though more or less straightforward on a conceptual level, can be nontrivial in practice. (Detailed discussions are in section 4.2.) Let $\partial D(Z)$, $Z \in \mathbb{R}^d$, $d \geq 1$, be the parameterization of the random boundary. A partial differential equation defined on this domain can be written as

$$\begin{cases} u_t(x, t) = \mathcal{L}(x; u), & D(Z) \times (0, T], \\ \mathcal{B}(u) = 0, & \partial D(Z) \times [0, T], \\ u = u_0, & D(Z) \times \{t = 0\}, \end{cases} \tag{8.1}$$

where $x = (x_1, \ldots, x_\ell)$, $\ell = 1, 2, 3$, is the coordinate in the random domain $D(Z)$. For simplicity, here the only source of uncertainty is assumed to be from $\partial D(Z)$. Note that even though the form of the governing equations is deterministic (it does not need to be), the solution still depends on the random variables Z and is a stochastic quantity. That is,

$$u(x, t) : D(Z) \times [0, T] \to \mathbb{R}^{n_u}$$

depends implicitly on the random variables $Z \in \mathbb{R}^d$.

The key to solving (8.1) is to define the problem on a fixed domain where the operations for statistical averaging become meaningful. A general approach is to use a one-to-one mapping ([130]). Let $y = (y_1, \ldots, y_\ell)$, $\ell = 1, 2, 3$, be the coordinate in a fixed domain $E \subset \mathbb{R}^\ell$ and let

$$y = y(x, Z), \qquad x = x(y, Z), \qquad \forall Z \in \mathbb{R}^d, \tag{8.2}$$

be a one-to-one mapping and its inverse such that the random domain $D(Z)$ can be uniquely transformed to the deterministic domain E. Then (8.1) can be transformed to the following: for all $Z \in \mathbb{R}^d$, find $v = v(y, Z) : \bar{E} \times \mathbb{R}^d \to \mathbb{R}^{n_u}$ such that

$$\begin{cases} v_t(y, t, Z) = L(y, Z; v), & E \times (0, T] \times \mathbb{R}^d, \\ B(v) = 0, & \partial E \times [0, T] \times \mathbb{R}^d, \\ v = v_0, & E \times \{t = 0\} \times \mathbb{R}^d, \end{cases} \tag{8.3}$$

where the operators L and B are transformed from $\mathcal{L}$ and $\mathcal{B}$, respectively, and v_0 is transformed from u_0 because of the random mapping (8.2). The transformed problem (8.3) is a stochastic PDE in a fixed domain, and the standard numerical techniques, including those based on gPC methodology, can be readily applied.

The mapping technique seeks to transform a problem defined in a random domain into a stochastic problem defined in a fixed domain. The randomness in the domain specification is absorbed into the mapping and further into the definition of the transformed equation. Thus, it is crucial to construct a unique and invertible mapping that is also robust and efficient in practice. For some domains, this can be achieved analytically ([102]).

Example 8.1 (Mapping for a random channel domain). Consider a straight channel in two dimensions, with length L and height H. Let us assume that the bottom boundary is a random process with zero mean value and other known distribution functions. That is, the channel is defined as

$$(x_1, x_2) \in D(\omega) = [0, L] \times [h(x, \omega), H], \tag{8.4}$$

where $\mathbb{E}[h(x, \omega)] = 0$. It is easy to see that a simple mapping

$$y_1 = x_1, \qquad y_2 = \frac{H}{H - h(x, \omega)}(x_2 - h(x, \omega))$$

can map the domain into

$$(y_1, y_2) \in E = [0, L] \times [0, H].$$

Example 8.2 (Mapping of a diffusion equation). Consider a deterministic Poisson's equation with homogeneous Dirichlet boundary conditions in a random domain $D(Z(\omega))$, $Z \in I_Z$,

$$\nabla \cdot [c(x)\nabla u(x, Z)] = a(x) \quad \text{in } D(Z),$$

$$u(x, Z) = 0 \qquad \text{on } \partial D(Z),$$

(8.5)

where no randomness exists in the diffusivity field $c(x)$ and the source field $a(x)$. The stochastic mapping (8.2) transforms (8.5) into a stochastic Poisson's equation in a deterministic domain E:

$$\sum_{i=1}^{\ell} \frac{\partial}{\partial y_i} \left[\kappa(y, Z) \sum_{j=1}^{\ell} \left(\alpha_{ij}(y, Z) \frac{\partial v}{\partial y_j} \right) \right] = J^{-1} f(y, Z) \quad \text{in } E \times I_Z, \quad (8.6)$$

$$v(y, Z) = 0 \qquad \text{on } \partial E \times I_Z,$$

where the random fields κ and f are the transformations of c and a, respectively, J is the transformation Jacobian

$$J(y, Z) = \frac{\partial(y_1, \ldots, y_\ell)}{\partial(x_1, \ldots, x_\ell)},$$

and

$$\alpha_{ij}(y, Z) = J^{-1} \nabla y_i \cdot \nabla y_j, \qquad 1 \le i, j \le \ell. \quad (8.7)$$

Though (8.6) is more complex, it is a stochastic diffusion problem in a fixed domain. The existing methods, such as those based on gPC, can be readily applied.

Example 8.3 (Diffusion in a random channel domain). Now let us combine the aforementioned two examples and consider diffusion problem in a random channel domain. This is the same example as that used in [130].

Consider the steady-state diffusion (8.5) with $a = 0$ and constant diffusivity $c(x)$ in a two-dimensional channel (8.4). To be specific, we set $L = 5$, $H = 1$, and the random bottom boundary as a random field with zero mean and an exponential two-point covariance function

$$C_{hh}(r, s) = \mathbb{E}[h(r, \omega)h(s, \omega)] = \sigma^2 \exp\left(-\frac{|r - s|}{b}\right), \qquad 0 \le r, s \le L, \quad (8.8)$$

where $b > 0$ is the correlation length. In the computational examples below, $b = L/5 = 1$, which corresponds to a boundary of moderate roughness. Finally, we prescribe Dirichlet boundary conditions $u = 1$ at $x_2 = H$ and $u = 0$ elsewhere.

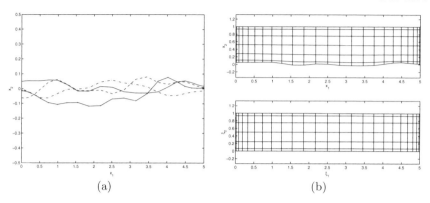

Figure 8.1 Channels with a rough wall generated with the 10-term ($K = 10$) KL expansion (8.9). (a) Four sample realizations of the bottom boundary $s(x_1, \omega_j)$ ($j = 1, \ldots, 4$). (b) A sample realization of the channel in the physical domain (x_1, x_2) and in the mapped domain in (ξ_1, ξ_2). Chebyshev meshes are used in both domains. (More details are in [130].)

We employ the finite-term Karhunen-Loève (KL)–type expansion (4.8) to decompose the boundary process. That is,

$$h(x_1, \omega) \approx \sigma \sum_{k=1}^{d} \sqrt{\lambda_k} \psi_k(x_1) Z_k(\omega), \qquad (8.9)$$

where $\{\lambda_k, \psi_k(x_1)\}$ are the eigenvalues and eigenfunctions of the integral equations

$$\int_0^L C_{hh}(r, s) \psi_k(r) dr = \lambda_k \psi_k(s), \qquad k = 1, \ldots, d. \qquad (8.10)$$

We further set $\{Z_i(\omega)\} \sim U(-1, 1)$ to be independent uniform random variables in $(-1, 1)$ and use the parameter $0 < \sigma < 1$ to control the maximum deviation of the randomness. (In the computational examples in this section, we set $\sigma = 0.1$.) We employ Legendre polynomials as the gPC basis functions.

It is worthwhile to stress again that the expansion (8.9) introduces two sources of errors—errors due to the finite d-term truncation and errors due to the assumption of independence of $\{Z_k(\omega)\}$. The truncation error is typically controlled by selecting the value of d to ensure that the eigenvalues $\{\lambda_k\}$ with $k > d$ are sufficiently small. For example, in this example the expansion with $d = 10$ captures 95 percent of the total spectrum (i.e., eigenvalues). A few realizations of the bottom boundary, obtained by the 10-term KL expansion, are shown in figure 8.1a. In figure 8.1b, one realization of the channel domain is mapped onto the corresponding rectangular domain $E = [0, L] \times [0, H]$. Also shown here are the Chebyshev collocation mesh points that are used to solve the mapped stochastic diffusion problem (8.6).

Figure 8.2 shows the first two moments of the solution, i.e., its mean (top) and standard deviation (STD) (bottom). We observe that the STD reaches its maximum close to, but not at, the random bottom boundary.

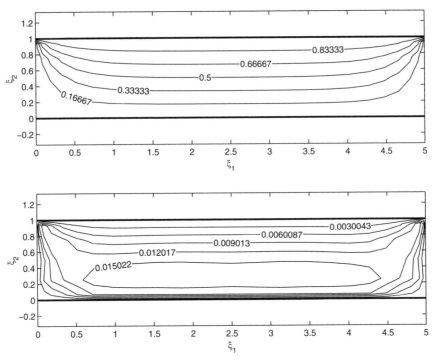

Figure 8.2 The mean and the STD of the dependent variable u computed with the stochastic Galerkin method.

To ascertain the convergence of the polynomial chaos expansion, we examine the STD profile along the cross section $y = 0.25$, where the STD is close to its maximum. Figure 8.3 shows the STD profiles obtained with different orders of Legendre expansions. One can see that the second order is sufficient for the Legendre expansion to converge. Although not shown here, the convergence of the mean solution is similar to that of the STD.

Monte Carlo simulations (MCSs) are also conducted to verify the results obtained by the stochastic Galerkin method. Figure 8.4 compares the STD profile along the cross section $y = 0.25$ computed via the second-order Legendre expansion with those obtained from MCS. We observe that as the number of realizations increases, the MCS results converge to the converged SG results. With about 2,000 realizations, the MCS results agree well with the SG results. In this case, at second order ($N = 2$) and 10 random dimensions ($d = 10$), the gPC stochastic Galerkin method requires $(N + d)!/N!d! = 66$ basis functions and is computationally more efficient than Monte Carlo simulation.

Often analytical mapping is not available; then a numerical technique can be employed to determine the mapping ([130]). This involves solving of a set of boundary value problems for the mapping. Other techniques for casting random domain problems into deterministic problems include the boundary perturbation method [126],

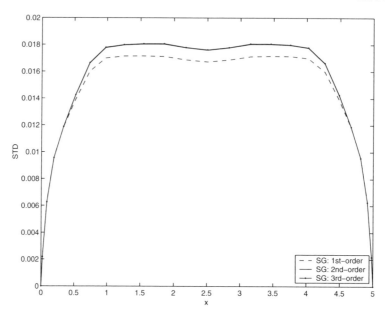

Figure 8.3 The STD profiles along the cross section $y = 0.25$ computed with the first-, second-, and third-degree Legendre polynomials.

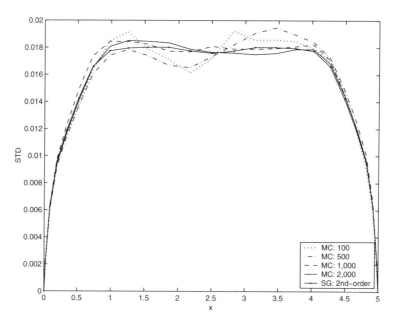

Figure 8.4 The STD profiles along the cross section $y = 0.25$ computed with the SG method (second-order Legendre chaos) and Monte MCS consisting of 100, 500, 1000, and 2000 realizations.

isoparametric mapping [15], the fictitious domain method [14], the eXtended finite element method [82], and a Lagrangian approach that works well for solid deformation [2]. Interested readers are strongly encouraged to consult the references.

8.2 BAYESIAN INVERSE APPROACH FOR PARAMETER ESTIMATION

When solving a stochastic system, e.g., (4.13), it is important that the probability distribution functions of the input random variables Z be available. This, along with the independence condition, allows us to sample the inputs or to build basis functions (such as gPC) to solve the system. A question naturally arises: What if there is not enough available information to determine and specify the parameter distributions? This is a problem of practical concern because in many cases there are not enough measurement data on the parameters—some parameters cannot even be measured. However, sometimes other measurement data are available—data not on the parameters but on some other quantities that can be computed. In this case, an inverse parameter estimation can be carried out to estimate the *true* distributions of the input random parameters.

The field of parameter estimation is not new—much research has gone into it for decades. Here, however, we will present an approach using Bayesian inference. More importantly, the approach is built upon the gPC algorithms in such way that it does not incur any simulation effort in addition to a one-time gPC simulation. In other words, for the inverse problem here, a forward gPC simulation is required only once, and the rest becomes off-line postprocessing. The strong approximation property of gPC expansion plays a vital role here.

Let us assume that each random variable Z_i has a *prior distribution* $F_i(z_i) = P(Z_i \leq z_i) \in [0, 1]$. The distribution can be made upon assumption, intuition, or even speculation when there are not sufficient data. Here we focus on continuous random variables. Subsequently, each Z_i has a probability density function $\pi_i(z_i) = dF_i(z_i)/dz_i$. We also assume the variables are mutually independent—another assumption when not enough data are available to suggest otherwise. Thus, the joint prior density function for Z is

$$\pi_Z(z) = \prod_{i=1}^{n_z} \pi_i(z_i). \tag{8.11}$$

Whenever possible, we will neglect the subscript of each probability density and use $\pi(z)$ to denote the probability density function of the random variable Z, $\pi_Z(z)$, unless confusion would arise.

Let

$$d^t = g(u) \in \mathbb{R}^{n_d} \tag{8.12}$$

be a set of variables that one observes, where $g : \mathbb{R}^{n_u} \to \mathbb{R}^{n_d}$ is a function relating the solution u to the true observable d^t. We then define a *forward model* $G : \mathbb{R}^{n_z} \to \mathbb{R}^{n_d}$ to describe the relation between the random parameters Z and the observable d^t:

$$d^t = G(Z) \triangleq g \circ u(Z). \tag{8.13}$$

In practice, measurement error is inevitable and the observed data d may not match the true value of d^t. Assuming additive observational errors, we have

$$d = d^t + e = G(Z) + e, \qquad (8.14)$$

where $e \in \mathbb{R}^{n_d}$ are mutually independent random variables with probability density functions $\pi(e) = \prod_{i=1}^{n_d} \pi(e_i)$. We make the usual assumption that e are also independent of Z.

The Bayesian approach seeks to estimate the parameters Z when given a set of observations d. To this end, Bayes' rule takes the form

$$\pi(z|d) = \frac{\pi(d|z)\pi(z)}{\int \pi(d|z)\pi(z)dz}, \qquad (8.15)$$

where $\pi(z)$ is the prior probability density of Z; $\pi(d|z)$ is the likelihood function; and $\pi(z|d)$, the density of Z conditioned on the data d, is the *posterior probability density* of Z. For notational convenience, we will use $\pi^d(z)$ to denote the posterior density $\pi(z|d)$ and $L(z)$ to denote the likelihood function $\pi(d|z)$. That is, (8.15) can be written as

$$\pi^d(z) = \frac{L(z)\pi(z)}{\int L(z)\pi(z)dz}. \qquad (8.16)$$

Following the independence assumption on the measurement noise e, the likelihood function is

$$L(z) \triangleq \pi(d|z) = \prod_{i=1}^{n_d} \pi_{e_i}(d_i - G_i(z)). \qquad (8.17)$$

The formulation, albeit simple, poses a challenge in practice. This is largely because the posterior distribution π^d does not have a closed and explicit form, thus preventing one from sampling it directly. A large amount of literature has been devoted to this challenge, with one of the most widely used approaches being the Markov chain Monte Carlo method (MCMC). For an extensive review, see [101]. Since most of the approaches are based on sampling, the main concern is to improve efficiency because each sampling point requires a solution of the underlying forward problem $G(Z)$ and can be time-consuming.

The gPC-based numerical method, in addition to its efficiency in solving the forward problem, provides another remarkable advantage here. For all $1 \le i \le n_d$, let $G_{N,i}(Z)$ be a gPC approximation for the ith component of $G(Z)$,

$$G_{N,i}(Z) = \sum_{|\mathbf{k}|=0}^{N} \hat{a}_{\mathbf{k},i} \Phi_{\mathbf{k}}(Z), \qquad (8.18)$$

where the expansion coefficients $\{\hat{a}_{\mathbf{k},i}\}$ are obtained by either a stochastic Galerkin or a stochastic collocation method. Then we effectively have *an analytical representation* of the forward problem in terms of Z, which can be sampled at an arbitrarily large number of nodes by simply evaluating the polynomial expression at the nodes. Thus, if we substitute the gPC approximation into the likelihood function, we obtain an approximate posterior density that can be easily sampled *without any simulation effort*, in addition to the stochastic forward problem which needs to be solved only once.

The gPC approximation of the posterior probability is

$$\pi_N^d(z) = \frac{L_N(z)\pi(z)}{\int L_N(z)\pi(z)dz}, \tag{8.19}$$

where $\pi(z)$ is again the prior density of Z and L_N is the approximate likelihood function defined as

$$L_N(z) \triangleq \pi_N(d|z) = \prod_{i=1}^{n_d} \pi_{e_i}(d_i - G_{N,i}(z)). \tag{8.20}$$

The error of the approximation can be quantified by using Kullback-Leibler divergence (KLD), which measures the difference between probability distributions and is defined, for probability density functions $\pi_1(z)$ and $\pi_2(z)$, as

$$D(\pi_1\|\pi_2) \triangleq \int \pi_1(z) \log \frac{\pi_1(z)}{\pi_2(z)} dz. \tag{8.21}$$

It is always nonnegative, and $D(\pi_1\|\pi_2) = 0$ when $\pi_1 = \pi_2$.

Under the common assumption that the observational error in (8.14) is independently and identically distributed (i.i.d.) Gaussian, e.g.,

$$e \sim N(0, \sigma^2 \mathbb{I}), \tag{8.22}$$

where $\sigma > 0$ is the standard deviation and $\mathbb{I}$ is an identity matrix of size $n_d \times n_d$, the following results can be established.

Lemma 8.4. *Assume that the observational error in (8.14) has an i.i.d. Gaussian distribution (8.22). If the gPC expansion $G_{N,i}$ in (8.18) of the forward model converges to G_i,*

$$\|G_i(Z) - G_{N,i}(Z)\|_{L^2_{\pi_Z}} \to 0, \qquad 1 \le i \le n_d, \qquad N \to \infty,$$

then the posterior probability π_N^d in (8.19) converges to the true posterior probability π^d in (8.16) in the sense that the Kullback-Leibler divergence (8.21) converges to zero; i.e.,

$$D(\pi_N^d\|\pi^d) \to 0, \qquad N \to \infty. \tag{8.23}$$

Theorem 8.5. *Assume that the observational error in (8.14) has an i.i.d. Gaussian distribution (8.22) and that the gPC expansion $G_{N,i}$ in (8.18) of the forward model converges to G_i in the form of*

$$\|G_i(Z) - G_{N,i}(Z)\|_{L^2_{\pi_Z}} \le C N^{-\alpha}, \qquad 1 \le i \le n_d,$$

where C is a constant independent of N and $\alpha > 0$, then for sufficiently large N,

$$D(\pi_N^d\|\pi^d) \lesssim N^{-\alpha}. \tag{8.24}$$

Proofs of these results can be found in [76].

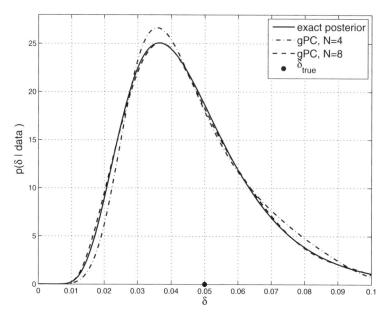

Figure 8.5 The posterior distribution of the boundary condition uncertainty δ (denoted as $p(\delta|data)$ here) of the Burgers' equation. The exact posterior from Bayes' rule is shown here along with the numerical results from the gPC collocation–based algorithms with order $N = 4$ and $N = 8$. Also shown is the true (and unknown) perturbation of δ^t.

Example 8.6 (Burgers' equation). Let us return to the Burgers' equation example in section 1.1.1. As shown in figure 1.1, a small amount of uncertainty with $\delta \sim \mathcal{U}(0, 0.1)$ can produce a large response in the location of the transition layer because of the supersensitive nature of the problem. On the other hand, the prediction produces the distribution range of the output, albeit correctly, too big to be useful. This suggests that the assumption about the uncertainty of δ is too wide and should be refined both in its range and distribution. This can be achieved using the Bayesian inverse estimation when some observation data are available.

Without actual experimental data, we generate "data" numerically. This is accomplished by fixing a "true" perturbation δ^t, conducting a high-order deterministic simulation to compute the true location of the transition layer d^t at steady state, and then adding Gaussian noise e to produce the data $d = d^t + e$. The data d are then used for the Bayesian inverse estimate for the posterior distribution of δ.

In figure 8.5, the numerical results of the posterior density are shown with gPC orders of $N = 4$ and $N = 8$. Compared to the exact posterior density obtained from the Bayesian rule directly (in this case the exact solution can be solved), we notice that the gPC results converge. For reference, the true and yet unknown location of the perturbation δ^t is also plotted. We observe that the posterior density clusters around the truth, as expected. The convergence of the gPC-based Bayesian algorithm is examined in more detail in figure 8.6 for orders as high as $N = 200$.

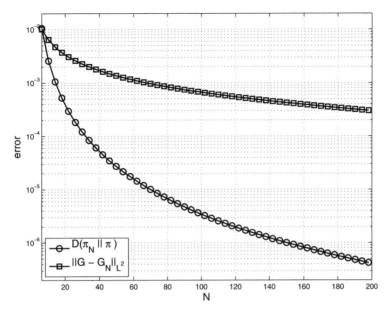

Figure 8.6 Convergence of the numerical posterior distribution by gPC-based Bayesian algorithms at increasing orders, along with the convergence of the gPC forward model prediction.

We observe the familiar exponential convergence. The convergence of the gPC forward problem solver is also plotted. It is clear that the posterior density converges at least as fast as (in fact, even faster than) the forward problem. This is consistent with the theoretical result. For more details on the analysis and numerical results, see [76].

8.3 DATA ASSIMILATION BY THE ENSEMBLE KALMAN FILTER

Any mathematical and numerical models, deterministic or stochastic, no matter how sophisticated, are approximations to the true physics. Though many models are accurate for a wide range of spatial and time domains, many can deviate from the "truth" quickly. (For example, weather forecasting.) In addition to improving the models, another way to improve the prediction is to take advantage of measurement data, which reflect the physical truth, sometimes partially (scarce and indirect measurement) and approximately (inaccurate measurement).

In data assimilation, data arrive sequentially in time, and one seeks to incorporate both the data and the prediction of the mathematical/numerical models to produce better predictions. There has been an extremely large amount of literature on various methods of data assimilation, with the most popular ones based on either a Kalman filter (KF) [54] or a particle filter. Here we discuss only the ensemble Kalman filter (EnKF) [28], a variance of the Kalman filter, and explain how gPC-based stochastic methods can be used to significantly improve the performance of the EnKF. The notation here is somewhat different from that in the rest of the book, particularly the use of bold letters, as we try to follow the notations commonly used in the data assimilation literature.

Let $\mathbf{u}^f \in \mathbb{R}^m$, $m \geq 1$, be a vector of *forecast* state variables (denoted by the superscript f) that are modeled by the following system:

$$\frac{d\mathbf{u}^f}{dt} = f(t, \mathbf{u}^f), \qquad t \in (0, T], \tag{8.25}$$

$$\mathbf{u}^f(0) = \mathbf{u}_0(Z), \tag{8.26}$$

where $T > 0$ and $Z \in \mathbb{R}^d$, $d \geq 1$, is a set of random variables parameterizing the random initial condition. The model (8.25) and (8.26) is obviously not a perfect model for the true physics, and the forecast may not represent the true state variables, $\mathbf{u}^t \in \mathbb{R}^m$, sufficiently well. If a set of *measurements* $\mathbf{d} \in \mathbb{R}^\ell$, $\ell \geq 1$, are available as

$$\mathbf{d} = \mathbf{H}\mathbf{u}^t + \boldsymbol{\epsilon}, \tag{8.27}$$

where $\mathbf{H} : \mathbb{R}^m \rightarrow \mathbb{R}^\ell$ is a measurement operator relating the *true state* variables $\mathbf{u}^t$ and the observation vector $\mathbf{d} \in \mathbb{R}^\ell$ and $\boldsymbol{\epsilon} \in \mathbb{R}^\ell$ is measurement error. Note that the measurement operator can be nonlinear, although it is written here in a linear fashion by following the traditional exposition of the (ensemble) Kalman filter. Also, characterization of the true state variables $\mathbf{u}^t$ can be highly nontrivial in practice. Here we assume that they are well-defined variables with dimension m.

The objective of data assimilation is to construct an optimal estimate of the true state, the *analyzed state* vector denoted as $\mathbf{u}^a \in \mathbb{R}^m$, based on the forecast $\mathbf{u}^f$ and the observation $\mathbf{d}$. Note that it is possible to add a noise term in (8.25) as a model for the modeling error. Here we restrict ourselves to the deterministic model (8.25) with the random initial condition (8.26).

8.3.1 The Kalman Filter and the Ensemble Kalman Filter

The Kalman filter is a sequential data assimilation method that consists of two stages at each time level when data are available—a forecast stage where the system (8.25) and (8.26) is solved and an analysis stage where the analyzed state $\mathbf{u}^a$ is obtained. Let $\mathbf{P}^f \in \mathbb{R}^{m \times m}$ be the covariance matrix of the forecast solution $\mathbf{u}^f$. The analyzed solution $\mathbf{u}^a$ in the standard KF is determined as a combination of the forecast solution $\mathbf{u}^f$ and the measurement $\mathbf{d}$ in the following manner:

$$\mathbf{u}^a = \mathbf{u}^f + \mathbf{K}(\mathbf{d} - \mathbf{H}\mathbf{u}^f), \tag{8.28}$$

where $\mathbf{K}$ is the *Kalman gain matrix* defined as

$$\mathbf{K} = \mathbf{P}^f \mathbf{H}^T (\mathbf{H}\mathbf{P}^f \mathbf{H}^T + \mathbf{R})^{-1}. \tag{8.29}$$

Here the superscript T denotes matrix transpose, and $\mathbf{R} \in \mathbb{R}^{\ell \times \ell}$ is the covariance of the measurement error $\boldsymbol{\epsilon}$. The covariance function of the analyzed state $\mathbf{u}^a$, $\mathbf{P}^a \in \mathbb{R}^{m \times m}$, is then obtained by

$$\mathbf{P}^a = (\mathbf{I} - \mathbf{K}\mathbf{H})\mathbf{P}^f(\mathbf{I} - \mathbf{K}\mathbf{H})^T + \mathbf{K}\mathbf{R}\mathbf{K}^T = (\mathbf{I} - \mathbf{K}\mathbf{H})\mathbf{P}^f, \tag{8.30}$$

where $\mathbf{I}$ is the identity matrix. When the system (8.25) is linear, the KF can be applied in a straightforward manner, as equations for evolution of the solution co-variance can be derived. For nonlinear systems, explicit derivation of the equations for the covariance function is not possible. Subsequently, various approximations such as the extended Kalman filter (EKF) were developed. Their applicability is limited to a certain degree depending on the approximation procedure. Further-more, in practical applications, forwarding the covariance functions (8.30) in time requires explicit storage and computation of $\mathbf{P}^f$, which scales as $O(m^2)$ and can be inefficient when the dimension of the model states, m, is large.

The ensemble Kalman filter (EnKF) ([11, 28]) overcomes the limitations of the Kalman filter by using an ensemble approximation of the random state solutions. Let

$$(\mathbf{u}^f)_i, \qquad i = 1, \ldots, M, \quad M > 1, \tag{8.31}$$

be an ensemble of the forecast state variables $\mathbf{u}^f$, where each ensemble member is indexed by the subscript $i = 1, \ldots, M$ and is obtained by solving the full nonlinear system (8.25). The analysis step for the EnKF consists of the following update performed on each of the model state ensemble members

$$(\mathbf{u}^a)_i = (\mathbf{u}^f)_i + \mathbf{K}_e((\mathbf{d})_i - \mathbf{H}(\mathbf{u}^f)_i), \qquad i = 1, \ldots, M, \tag{8.32}$$

where

$$\mathbf{K}_e = \mathbf{P}_e^f \mathbf{H}^T (\mathbf{H}\mathbf{P}_e^f \mathbf{H}^T + \mathbf{R}_e)^{-1} \tag{8.33}$$

is the *ensemble Kalman gain matrix*. Here

$$\begin{aligned} \mathbf{P}_e^f &\triangleq \overline{(\mathbf{u}^f - \overline{\mathbf{u}}^f)(\mathbf{u}^f - \overline{\mathbf{u}}^f)^T} \simeq \mathbf{P}^f, \\ \mathbf{P}_e^a &\triangleq \overline{(\mathbf{u}^a - \overline{\mathbf{u}}^a)(\mathbf{u}^a - \overline{\mathbf{u}}^a)^T} \simeq \mathbf{P}^a, \end{aligned} \tag{8.34}$$

are the approximate forecast covariance and analysis covariance, respectively, obtained by using statistical averages of the solution ensemble (denoted by the overbar), and $\mathbf{R}_e = \overline{\boldsymbol{\epsilon}\boldsymbol{\epsilon}^T} \simeq \mathbf{R}$ is the approximate observation error covariance. Therefore, the covariance functions are approximated by ensemble averages and do not need to be forwarded in time explicitly. An extensive review of the EnKF can be found in [29].

8.3.2 Error Bound of the EnKF

As an approximation to the Kalman filter, one obvious source of error for the EnKF is from the sampling. Note that here we define error as *numerical error*—it is the difference between the result obtained by the KF and that obtained by the EnKF. This is different from *modeling error* between the result of the KF and the physi-cal truth. For EnKF, the overall error consists of both the numerical error and the modeling error. Here we study only the former.

To understand the impact of numerical error more precisely, we cite here an error bound of the EnKF derived in [66]. Let $t_1 < t_2 < \cdots$ be discrete time instances at which data arrive sequentially and assimilation is made. Without loss of gen-erality, let us assume that they are uniformly distributed with a constant step size

$\Delta T = t_k - t_{k-1}, \forall k > 1$. Let E_n be the numerical error of the EnKF, that is, the difference between the EnKF results and the exact KF results measured in a proper norm at time level t_n, $n \geq 1$, then the following bound holds:

$$E_n \leq \left(E_0 + \sum_{k=1}^{n} e_k \right) \exp(\Lambda \cdot t_n), \qquad (8.35)$$

where E_0 is the error of sampling the initial state, e_k is the local error at time level t_k, $1 \leq k \leq n$, and $\Lambda > 0$ is a constant. The local error scales as

$$e_k \sim O\left(\Delta t^p, \sigma M^{-\alpha} \right), \qquad \Delta t \to 0, M \to \infty, \qquad (8.36)$$

where $O(\Delta t^p)$ denotes the numerical integration error in time induced by solving (8.25) and (8.26) with a time step Δt and a temporal integration order $p \geq 1$, $\sigma > 0$ is the noise level of the measurement data and scales with the standard deviation of the measurement noise, M is the size of the ensemble, and $\alpha > 0$ is the convergence rate of the sampling scheme. For Monte Carlo sampling, $\alpha = 1/2$. In most cases, this sampling error dominates. A notable result is that the constant Λ depends on the size of the assimilation step in an inverse manner, i.e., $\Lambda \propto \Delta T^{-1}$. This implies that more frequent data assimilation by the EnKF can magnify the numerical errors. Since more frequent assimilation is always desirable (whenever data are available) for a better estimate of the true state, it is imperative to keep the numerical errors, particularly the sampling errors, of the EnKF under control. Although the sampling errors can be easily reduced by increasing the ensemble size, in practice this can significantly increase the computational burden, especially for large-scale problems.

8.3.3 Improved EnKF via gPC Methods

Here we demonstrate that one can yet again take advantage of a highly accurate gPC approximation to construct an improved EnKF scheme with much reduced numerical error. Let

$$\mathbf{u}_N^f(t, Z) = \sum_{|\mathbf{i}|=0}^{N} \hat{\mathbf{u}}_{\mathbf{i}}^f(t) \Phi_{\mathbf{i}}(Z) \qquad (8.37)$$

be the gPC solution to the forecast equations (8.25) and (8.26) with sufficiently high accuracy, where the expansion coefficients $\hat{\mathbf{u}}_{\mathbf{i}}^f(t)$ can be obtained by either the gPC Galerkin procedure or the gPC collocation procedure.

In addition to offering efficiency for the forecast solution, another (often overlooked) advantage of gPC expansion is that it provides an analytical representation of the solution in terms of the random inputs. All statistical information about $\mathbf{u}_N^f$ can be obtained analytically or with minimum computational effort. For example, the mean and covariance are

$$\overline{\mathbf{u}}_N^f = \hat{\mathbf{u}}_0^f, \qquad \mathbf{P}_N^f = \sum_{0 < |\mathbf{i}| \leq N} \left[\hat{\mathbf{u}}_{\mathbf{i}}^f \left(\hat{\mathbf{u}}_{\mathbf{i}}^f \right)^T \gamma_{\mathbf{i}} \right], \qquad (8.38)$$

respectively, where $\gamma_{\mathbf{i}} = \mathbb{E}[\Phi_{\mathbf{i}}^2]$. And they can be used as accurate approximations of the exact mean and covariance of the forecast solution $\mathbf{u}^f$. Furthermore, one can generate an ensemble of solution realizations by sampling the random variables Z in (8.37). This procedure involves nothing but polynomial evaluations, and thus generating an ensemble with an arbitrarily large number of samples does not require any computation of the original governing equations (8.25) and (8.26). Let

$$(\mathbf{u}_N^f)_i = \sum_{|\mathbf{k}|=0}^{N} \hat{\mathbf{u}}_{\mathbf{k}}^f(t) \Phi_{\mathbf{k}}((Z)_i), \qquad i = 1, \ldots, M, \quad M \gg 1, \tag{8.39}$$

be an ensemble of the forecast solution realizations with size M, where $(Z)_i, i = 1, \ldots, M$, are Monte Carlo samples of the random vector Z. Equipped with a knowledge of the solution statistics, particularly the mean and covariance from (8.38), we can apply the EnKF scheme (8.32) to obtain analyzed states.

$$(\mathbf{u}_N^a)_i = (\mathbf{u}_N^f)_i + \mathbf{K}_N(\mathbf{d}_i - \mathbf{H}(\mathbf{u}_N^f)_i), \qquad i = 1, \ldots, M, \tag{8.40}$$

where $\mathbf{K}_N$ is the *gPC Kalman gain matrix* defined as

$$\mathbf{K}_N = \mathbf{P}_N^f \mathbf{H}^T (\mathbf{H} \mathbf{P}_N^f \mathbf{H}^T + \mathbf{R})^{-1}, \tag{8.41}$$

which approximates the Kalman gain matrix (8.29).

The key ingredients and advantages of the gPC-based EnKF are

- Solution of the forecast problem by a gPC-based method, either a Galerkin or a collocation method. This, in many cases, is (much) more efficient than the traditional Monte Carlo sampling employed in the EnKF.
- At the update stage, one can use the gPC forward problem solution to generate an arbitrarily large number of samples and update them individually by (8.40), similar to the traditional EnKF. The ability to update a large ensemble of solutions results in a significant reduction in sampling errors. Moreover, this step is virtually "free," in the sense that generating the large solution ensemble via gPC is nothing but evaluation of the gPC polynomial expression and induces no simulation cost.
- For a linear system of equations (8.25) with Gaussian noise, the gPC-based EnKF becomes equivalent to the standard Kalman filter.

More details about these methods can be found in [67].

Appendix A

Some Important Orthogonal Polynomials
in the Askey Scheme

Here we summarize the definitions and properties of some important orthogonal polynomials from the Askey scheme. Denote $\{Q_n(x)\}$ as an orthogonal polynomial system with the orthogonal relation

$$\int_S Q_n(x)Q_m(x)w(x)dx = h_n^2 \delta_{mn}$$

for continuous x, or in the discrete case,

$$\sum_x Q_n(x)Q_m(x)w(x) = h_n^2 \delta_{mn},$$

where S is the support of $w(x)$. The three-term recurrence relation takes the form

$$-x Q_n(x) = b_n Q_{n+1}(x) + \gamma_n Q_n(x) + c_n Q_{n-1}(x), \qquad n \geq 0,$$

with initial conditions $Q_{-1}(x) = 0$ and $Q_0(x) = 1$. Another way of expressing the recurrence relation is

$$Q_{n+1}(x) = (A_n x + B_n)Q_n(x) - C_n Q_{n-1}(x), \qquad n \geq 0, \qquad \text{(A.1)}$$

where $A_n, C_n \neq 0$ and $C_n A_n A_{n-1} > 0$. It is straightforward to show that, if we scale variable x by denoting $y = \alpha x$ for $\alpha > 0$, then the recurrence relation takes the form

$$S_{n+1}(y) = (A_n y + \alpha B_n)S_n(y) - \alpha^2 C_n S_{n-1}(y). \qquad \text{(A.2)}$$

Another important property is that these orthogonal polynomials are solutions of a differential equation

$$s(x)y'' + \tau(x)y' + \lambda y = 0 \qquad \text{(A.3)}$$

in continuous cases, and a difference equation

$$s(x)\Delta\nabla y(x) + \tau(x)\Delta y(x) + \lambda y(x) = 0 \qquad \text{(A.4)}$$

in discrete cases, where $s(x)$ and $\tau(x)$ are polynomials of at most second and first degree, respectively, and λ is a constant. The notations for the discrete cases are

$$\Delta f(x) = f(x+1) - f(x), \qquad \nabla f(x) = f(x) - f(x-1).$$

When

$$\lambda = \lambda_n = -n\tau' - \frac{1}{2}n(n-1)s'',$$

the equations have a particular solution of the form $y(x) = Q_n(x)$, which is a polynomial of degree n.

A.1 CONTINUOUS POLYNOMIALS

A.1.1 Hermite Polynomial $H_n(x)$ and Gaussian Distribution

Definition:

$$H_n(x) = (2x)^n \, _2F_0\left(-\frac{n}{2}, -\frac{n-1}{2}; \; ; -\frac{2}{x^2}\right). \tag{A.5}$$

Orthogonality:

$$\int_{-\infty}^{\infty} H_m(x) H_n(x) w(x) dx = n! \delta_{mn}, \tag{A.6}$$

where

$$w(x) = \frac{1}{\sqrt{2\pi}} e^{-x^2/2}. \tag{A.7}$$

Recurrence relation:

$$H_{n+1}(x) = x H_n(x) - n H_{n-1}(x). \tag{A.8}$$

Rodriguez formula:

$$e^{-x^2/2} H_n(x) = (-1)^n \frac{d^n}{dx^n}\left(e^{-x^2/2}\right). \tag{A.9}$$

Differential equation:

$$y''(x) - x y'(x) + n y(x) = 0, \qquad y(x) = H_n(x). \tag{A.10}$$

A.1.2 Laguerre Polynomial $L_n^{(\alpha)}(x)$ and Gamma Distribution

Definition:

$$L_n^{(\alpha)}(x) = \frac{(\alpha+1)_n}{n!} \, _1F_1(-n; \alpha+1; x). \tag{A.11}$$

Orthogonality:

$$\int_0^{\infty} L_m^{(\alpha)}(x) L_n^{(\alpha)}(x) w(x) dx = \frac{(\alpha+1)_n}{n!} \delta_{mn}, \qquad \alpha > -1, \tag{A.12}$$

where

$$w(x) = \frac{x^\alpha e^{-x}}{\Gamma(\alpha+1)}. \tag{A.13}$$

Recurrence relation:

$$(n+1)L_{n+1}^{(\alpha)}(x) - (2n+\alpha+1-x)L_n^{(\alpha)}(x) + (n+\alpha)L_{n-1}^{(\alpha)}(x) = 0. \tag{A.14}$$

Normalized recurrence relation:

$$x q_n(x) = q_{n+1}(x) + (2n+\alpha+1)q_n(x) + n(n+\alpha)q_{n-1}(x), \tag{A.15}$$

where

$$L_n^{(\alpha)}(x) = \frac{(-1)^n}{n!} q_n(x).$$

Rodriguez formula:

$$e^{-x} x^\alpha L_n^{(\alpha)}(x) = \frac{1}{n!} \frac{d^n}{dx^n} \left(e^{-x} x^{n+\alpha} \right). \tag{A.16}$$

Differential equation:

$$xy''(x) + (\alpha + 1 - x)y'(x) + ny(x) = 0, \qquad y(x) = L_n^{(\alpha)}(x). \tag{A.17}$$

Recall that the gamma distribution has the probability density function

$$f(x) = \frac{x^\alpha e^{-x/\beta}}{\beta^{\alpha+1}\Gamma(\alpha+1)}, \qquad \alpha > -1, \ \beta > 0. \tag{A.18}$$

The weighting function of Laguerre polynomial (A.13) is the same as that of the gamma distribution with the scale parameter $\beta = 1$.

A.1.3 Jacobi Polynomial $P_n^{(\alpha,\beta)}(x)$ and Beta Distribution

Definition:

$$P_n^{(\alpha,\beta)}(x) = \frac{(\alpha+1)_n}{n!} \, {}_2F_1 \left(-n, n+\alpha+\beta+1; \alpha+1; \frac{1-x}{2} \right). \tag{A.19}$$

Orthogonality:

$$\int_{-1}^{1} P_m^{(\alpha,\beta)}(x) P_n^{(\alpha,\beta)}(x) w(x) dx = h_n^2 \delta_{mn}, \qquad \alpha, \beta > -1, \tag{A.20}$$

where

$$h_n^2 = \frac{(\alpha+1)_n(\beta+1)_n}{n!(2n+\alpha+\beta+1)(\alpha+\beta+2)_{n-1}},$$

$$w(x) = \frac{\Gamma(\alpha+\beta+2)}{2^{\alpha+\beta+1}\Gamma(\alpha+1)\Gamma(\beta+1)} (1-x)^\alpha (1+x)^\beta. \tag{A.21}$$

Recurrence relation:

$$x P_n^{(\alpha,\beta)}(x) = \frac{2(n+1)(n+\alpha+\beta+1)}{(2n+\alpha+\beta+1)(2n+\alpha+\beta+2)} P_{n+1}^{(\alpha,\beta)}(x)$$

$$+ \frac{\beta^2 - \alpha^2}{(2n+\alpha+\beta)(2n+\alpha+\beta+2)} P_n^{(\alpha,\beta)}(x)$$

$$+ \frac{2(n+\alpha)(n+\beta)}{(2n+\alpha+\beta)(2n+\alpha+\beta+1)} P_{n-1}^{(\alpha,\beta)}(x). \tag{A.22}$$

Normalized recurrence relation:

$$x p_n(x) = p_{n+1}(x) + \frac{\beta^2 - \alpha^2}{(2n+\alpha+\beta)(2n+\alpha+\beta+2)} p_n(x)$$

$$+ \frac{4n(n+\alpha)(n+\beta)(n+\alpha+\beta)}{(2n+\alpha+\beta-1)(2n+\alpha+\beta)^2(2n+\alpha+\beta+1)} p_{n-1}(x), \tag{A.23}$$

where

$$P_n^{(\alpha,\beta)}(x) = \frac{(n+\alpha+\beta+1)_n}{2^n n!} p_n(x).$$

Rodriguez formula:

$$(1 - x)^\alpha (1 + x)^\beta P_n^{(\alpha,\beta)}(x) = \frac{(-1)^n}{2^n n!} \frac{d^n}{dx^n} \left[(1 - x)^{n+\alpha} (1 + x)^{n+\beta} \right]. \quad \text{(A.24)}$$

Differential equation:

$$(1 - x^2) y''(x) + [\beta - \alpha - (\alpha + \beta + 2)x] y'(x) + n(n + \alpha + \beta + 1) y(x) = 0, \quad \text{(A.25)}$$

where $y(x) = P_n^{(\alpha,\beta)}(x)$.

A.2 DISCRETE POLYNOMIALS

A.2.1 Charlier Polynomial $C_n(x; a)$ and Poisson Distribution

Definition:

$$C_n(x; a) = {}_2F_0 \left(-n, -x; \ ; -\frac{1}{a} \right). \quad \text{(A.26)}$$

Orthogonality:

$$\sum_{x=0}^{\infty} \frac{a^x}{x!} C_m(x; a) C_n(x; a) = a^{-n} e^a n! \delta_{mn}, \qquad a > 0. \quad \text{(A.27)}$$

Recurrence relation:

$$-x C_n(x; a) = a C_{n+1}(x; a) - (n + a) C_n(x; a) + n C_{n-1}(x; a). \quad \text{(A.28)}$$

Rodriguez formula:

$$\frac{a^x}{x!} C_n(x; a) = \nabla^n \left(\frac{a^x}{x!} \right). \quad \text{(A.29)}$$

Difference equation:

$$-n y(x) = a y(x + 1) - (x + a) y(x) + x y(x - 1), \quad y(x) = C_n(x; a). \quad \text{(A.30)}$$

The probability function of *Poisson* distribution is

$$f(x; a) = e^{-a} \frac{a^x}{x!}, \qquad k = 0, 1, 2, \ldots. \quad \text{(A.31)}$$

Despite a constant factor e^{-a}, it is the same as the weighting function of Charlier polynomials.

A.2.2 Krawtchouk Polynomial $K_n(x; p, N)$ and Binomial Distribution

Definition:

$$K_n(x; p, N) = {}_2F_1 \left(-n, -x; -N; \frac{1}{p} \right), \qquad n = 0, 1, \ldots, N. \quad \text{(A.32)}$$

Orthogonality:

$$\sum_{x=0}^{N} \binom{N}{x} p^x (1-p)^{N-x} K_m(x; p, N) K_n(x; p, N)$$

$$= \frac{(-1)^n n!}{(-N)_n} \left(\frac{1-p}{p}\right)^n \delta_{mn}, \quad 0 < p < 1. \tag{A.33}$$

Recurrence relation:

$$-x K(x; p, N) = p(N-n) K_{n+1}(x; p, N) - [p(N-n) + n(1-p)] K_n(x; p, N)$$
$$+ n(1-p) K_{n-1}(x; p, N). \tag{A.34}$$

Rodriguez formula:

$$\binom{N}{x} \left(\frac{p}{1-p}\right)^x K_n(x; p, N) = \nabla^n \left[\binom{N-n}{x} \left(\frac{p}{1-p}\right)^x\right]. \tag{A.35}$$

Difference equation:

$$-ny(x) = p(N-x)y(x+1) - [p(N-x) - xq]y(x) + xq y(x-1), \tag{A.36}$$

where $y(x) = K_n(x; p, N)$ and $q = 1 - p$.

Clearly, the weighting function from (A.33) is the probability function of the *binomial* distribution.

A.2.3 Meixner Polynomial $M_n(x; \beta, c)$ and Negative Binomial Distribution

Definition:

$$M_n(x; \beta, c) = {}_2F_1\left(-n, -x; \beta; 1 - \frac{1}{c}\right). \tag{A.37}$$

Orthogonality:

$$\sum_{x=0}^{\infty} \frac{(\beta)_x}{x!} c^x M_m(x; \beta, c) M_n(x; \beta, c) = \frac{c^{-n} n!}{(\beta)_n (1-c)^\beta} \delta_{mn}, \quad \beta > 0, \ 0 < c < 1. \tag{A.38}$$

Recurrence relation:

$$(c-1)x M_n(x; \beta, c) = c(n+\beta) M_{n+1}(x; \beta, c) - [n + (n+\beta)c] M_n(x; \beta, c)$$
$$+ n M_{n-1}(x; \beta, c). \tag{A.39}$$

Rodriguez formula:

$$\frac{(\beta)_x c^x}{x!} M_n(x; \beta, c) = \nabla^n \left[\frac{(\beta + n)_x c^x}{x!}\right]. \tag{A.40}$$

Difference equation:

$$n(c-1)y(x) = c(x+\beta)y(x+1) - [x + (x+\beta)c]y(x) + xy(x-1), \tag{A.41}$$

where $y(x) = M_n(x; \beta, c)$.

The weighting function is

$$f(x) = \frac{(\beta)_x}{x!}(1-c)^\beta c^x, \qquad 0 < c < 1, \quad \beta > 0, \quad x = 0, 1, 2, \ldots. \qquad (A.42)$$

It can be verified that it is the probability function of a *negative binomial* distribution. In the case where β is an integer, it is often called the *Pascal* distribution.

A.2.4 Hahn Polynomial $Q_n(x; \alpha, \beta, N)$ and Hypergeometric Distribution

Definition:

$$Q_n(x; \alpha, \beta, N) = {}_3F_2(-n, n+\alpha+\beta+1, -x; \alpha+1, -N; 1), \quad n = 0, 1, \ldots, N. \qquad (A.43)$$

Orthogonality: For $\alpha > -1$ and $\beta > -1$ or for $\alpha < -N$ and $\beta < -N$,

$$\sum_{x=0}^{N} \binom{\alpha+x}{x}\binom{\beta+N-x}{N-x} Q_m(x; \alpha, \beta, N) Q_n(x; \alpha, \beta, N) = h_n^2 \delta_{mn}, \qquad (A.44)$$

where

$$h_n^2 = \frac{(-1)^n (n+\alpha+\beta+1)_{N+1}(\beta+1)_n n!}{(2n+\alpha+\beta+1)(\alpha+1)_n(-N)_n N!}.$$

Recurrence relation:

$$-x Q_n(x) = A_n Q_{n+1}(x) - (A_n + C_n) Q_n(x) + C_n Q_{n-1}(x), \qquad (A.45)$$

where

$$Q_n(x) := Q_n(x; \alpha, \beta, N)$$

and

$$\begin{cases} A_n = \dfrac{(n+\alpha+\beta+1)(n+\alpha+1)(N-n)}{(2n+\alpha+\beta+1)(2n+\alpha+\beta+2)} \\[2ex] C_n = \dfrac{n(n+\alpha+\beta+N+1)(n+\beta)}{(2n+\alpha+\beta)(2n+\alpha+\beta+1)}. \end{cases}$$

Rodriguez formula:

$$w(x; \alpha, \beta, N) Q_n(x; \alpha, \beta, N) = \frac{(-1)^n(\beta+1)_n}{(-N)_n} \nabla^n [w(x; \alpha+n, \beta+n, N-n)], \qquad (A.46)$$

where

$$w(x; \alpha, \beta, N) = \binom{\alpha+x}{x}\binom{\beta+N-x}{N-x}.$$

Difference equation:

$$n(n+\alpha+\beta+1)y(x) = B(x)y(x+1) - [B(x)+D(x)]y(x) + D(x)y(x-1), \qquad (A.47)$$

where $y(x) = Q_n(x; \alpha, \beta, N)$, $B(x) = (x+\alpha+1)(x-N)$, and $D(x) = x(x-\beta-N-1)$.

If we set $\alpha = -\tilde{\alpha} - 1$ and $\beta = -\tilde{\beta} - 1$, we obtain

$$\tilde{w}(x) = \frac{1}{\binom{N-\tilde{\alpha}-\tilde{\beta}-1}{N}} \frac{\binom{\tilde{\alpha}}{x}\binom{\tilde{\beta}}{N-x}}{\binom{\tilde{\alpha}+\tilde{\beta}}{N}}.$$

Apart from the constant factor $1/\binom{N-\tilde{\alpha}-\tilde{\beta}-1}{N}$, this is the definition of *hypergeometric distribution*.

Appendix B

The Truncated Gaussian Model $G(\alpha, \beta)$

The truncated Gaussian model was developed in [123] in order to circumvent the mathematical difficulty resulting from the tails of Gaussian distribution. It is an approximation of Gaussian distribution by generalized polynomial chaos (gPC) Jacobi expansion. The approximation can be improved either by increasing the order of the gPC expansion or by adjusting the parameters in the Jacobi polynomials. The important property of the model is that it has bounded support, i.e., no tails. This can be used as an alternative in practical applications where the random inputs resemble Gaussian distribution *and* the boundedness of the supports is critical to the solution procedure.

While the procedure of approximating (weakly) a Gaussian distribution by Jacobi polynomials was explained in section 5.1.2, here we tabulate the results for future reference. The gPC Jacobi approximation for Gaussian distribution is denoted as $G(\alpha, \beta)$ with $\alpha, \beta > -1$. Because of the symmetry of Gaussian distribution, we set $\alpha = \beta$ in the Jacobi polynomials.

In figures B.1–B.3, the probability density functions (PDFs) of the gPC Jacobi chaos approximations are plotted for values of $\alpha = \beta = 0$ to 10. For $\alpha = \beta = 0$, Jacobi chaos becomes Legendre chaos, and the first-order expansion is simply a uniform random variable. In this case, Gibbs' oscillations are observed. As the values of (α, β) increase, the approximations improve. The expansion coefficients at different orders are tabulated in table B.1, together with the errors in variance and kurtosis compared with the exact Gaussian distribution. It is seen that, with

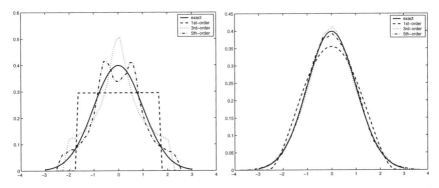

Figure B.1 Gaussian random variables approximated by Jacobi chaos. Left: $\alpha = \beta = 0$. Right: $\alpha = \beta = 2$.

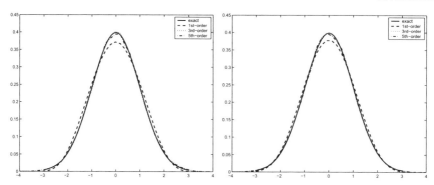

Figure B.2 Gaussian random variables approximated by Jacobi chaos. Left: $\alpha = \beta = 4$.
Right: $\alpha = \beta = 6$.

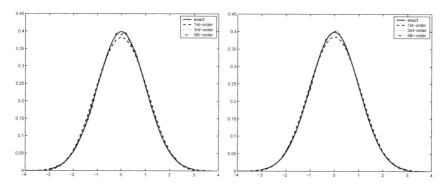

Figure B.3 Gaussian random variables approximated by Jacobi chaos. Left: $\alpha = \beta = 8$.
Right: $\alpha = \beta = 10$.

$\alpha = \beta = 10$, even the first-order approximation, which is simply a beta random variable, has an error in variance of as little as 0.1 percent. The errors in kurtosis are larger because the Jacobi chaos approximations do not possess tails. This, however, is exactly our objective.

Table B.1 Approximating Gaussian Random Variables via Jacobi Chaos: Expansion Coefficients y_k and errors[a]

	$\alpha = \beta = 0$	$\alpha = \beta = 2$	$\alpha = \beta = 4$	$\alpha = \beta = 6$	$\alpha = \beta = 8$	$\alpha = \beta = 10$
y_1	1.69248	8.7827(−1)	6.6218(−1)	5.5273(−1)	4.8399(−1)	4.3575(−1)
ϵ_2	4.51704(−2)	8.25346(−3)	3.46301(−3)	2.00729(−3)	1.38842(−3)	1.07231(−3)
ϵ_4	1.35894	7.05024(−1)	4.79089(−1)	3.63557(−1)	2.93246(−1)	2.45916(−1)
y_3	4.8399(−1)	7.5493(−2)	2.6011(−2)	1.2216(−2)	6.77970(−3)	4.17792(−3)
ϵ_2	1.17071(−2)	8.51816(−4)	4.49245(−4)	4.23983(−4)	4.33894(−4)	4.45282(−4)
ϵ_4	5.02097(−1)	7.97474(−2)	3.33201(−2)	2.40064(−2)	2.21484(−2)	2.22539(−2)
y_5	2.7064(−1)	1.9959(−2)	2.9936(−3)	2.3531(−4)	−3.30888(−4)	−4.19539(−4)
ϵ_2	5.04838(−3)	3.97059(−4)	3.96880(−4)	4.22903(−4)	4.28283(−4)	4.25043(−4)
ϵ_4	2.55526(−1)	2.29373(−2)	1.92101(−2)	2.15095(−2)	2.06846(−2)	2.08317(−2)

[a] ϵ_2 is the error in variance; ϵ_4 is the error in kurtosis. There is no error in the mean. $y_k = 0$ when k is even.

References

[1] S. Acharjee and N. Zabaras. Uncertainty propagation in finite deformations—a spectral stochastic Lagrangian approach. *Comput. Methods Appl. Mech. Engrg.*, 195:2289–2312, 2006.

[2] N. Agarwal and N.R. Aluru. A stochastic Lagrangian approach for geometrical uncertainties in electrostatics. *J. Comput. Phys.*, 226(1):156–179, 2007.

[3] N. Agarwal and N.R. Aluru. A domain adaptive stochastic collocation approach for analysis of MEMS under uncertainties. *J. Comput. Phys.*, 228(20):7662–7688, 2009.

[4] T.W. Anderson. *An Introduction to Multivariate Statistical Analysis*. John Wiley & Sons, New York, 1958.

[5] B.V. Asokan and N. Zabaras. A stochastic variational multiscale method for diffusion in heterogeneous random media. *J. Comput. Phys.*, 218:654–676, 2006.

[6] K. Atkinson and W. Han. *Theoretical Numerical Analysis*. Springer-Verlag, New York, 2001.

[7] I. Babuška, F. Nobile, and R. Tempone. A stochastic collocation method for elliptic partial differential equations with random input data. *SIAM J. Numer. Anal.*, 45(3):1005–1034, 2007.

[8] I. Babuška, R. Tempone, and G.E. Zouraris. Galerkin finite element approximations of stochastic elliptic differential equations. *SIAM J. Numer. Anal.*, 42:800–825, 2004.

[9] V. Barthelmann, E. Novak, and K. Ritter. High dimensional polynomial interpolation on sparse grids. *Adv. Comput. Math.*, 12:273–288, 1999.

[10] P. Beckmann. *Orthogonal Polynomials for Engineers and Physicists*. Golem Press, Boulder, Colorado, 1973.

[11] G. Burgers, P.V. Leeuwen, and G. Evensen. Analysis scheme in the ensemble Kalman filter. *Mon. Weather Rev.*, 126:1719–1724, 1998.

[12] R.H. Cameron and W.T. Martin. The orthogonal development of nonlinear functionals in a series of Fourier-Hermite functionals. *Ann. of Math.*, 48:385–392, 1947.

[13] C. Canuto, M.Y. Hussaini, A. Quarteroni, and T.A. Zang. *Spectral Methods in Fluid Dynamics*. Springer-Verlag, New York, 1988.

[14] C. Canuto and T. Kozubek. A fictitious domain approach to the numerical solutions of PDEs in stochastic domains. *Numer. Math.*, 107(2):257–293, 2007.

[15] C. Chauviere, J.S. Hesthaven, and L. Lurati. Computational modeling of uncertainty in time-domain electromagnetics. *SIAM J. Sci. Comput.*, 28:751–775, 2006.

[16] C. Chauviere, J.S. Hesthaven, and L. Wilcox. Efficient computation of RCS from scatterers of uncertain shapes. *IEEE Trans. Antennas and Propagation*, 55(5):1437–1448, 2007.

[17] Q.-Y. Chen, D. Gottlieb, and J.S. Hesthaven. Uncertainty analysis for the steady-state flows in a dual throat nozzle. *J. Comput. Phys.*, 204:387–398, 2005.

[18] E.W. Cheney. *Introduction to Approximation Theory*. McGraw-Hill, New York, 1966.

[19] T.S. Chihara. *An Introduction to Orthogonal Polynomials*. Gordon and Breach, New York, 1978.

[20] A.J. Chorin. Gaussian fields and random flow. *J. Fluid Mech.*, 85:325–347, 1974.

[21] R. Cools. An encyclopaedia of cubature formulas. *J. Complexity*, 19:445–453, 2003.

[22] R. Courant and D. Hilbert. *Methods of Mathematical Physics*. John Wiley & Sons, New York, 1953.

[23] G. Deodatis. Weighted integral method. I. Stochastic stiffness matrix. *J. Engrg. Math.*, 117(8):1851–1864, 1991.

[24] G. Deodatis and M. Shinozuka. Weighted integral method. II. Response variability and reliability. *J. Engrg. Math.*, 117(8):1865–1877, 1991.

[25] L. Devroye. *Non-uniform Random Variate Generation*. Springer-Verlag, New York, 1986.

[26] A. Doostan, R.G. Ghanem, and J. Red-Horse. Stochastic model reduction for chaos representations. *Comput. Methods Appl. Mech. Engrg.*, 196:3951–3966, 2007.

[27] H. Engels. *Numerical Quadrature and Cubature*. Academic Press, London, 1980.

[28] G. Evensen. Sequential data assimilation with a nonlinear quasi-geostrophic model using Monte Carlo methods to forecast error statistics. *J. Geophys. Res.*, 99:10,143–10,162, 1994.

[29] G. Evensen. *Data Assimilation: The Ensemble Kalman Filter*. Springer-Verlag, Berlin, 2007.

[30] G.S. Fishman. *Monte Carlo: Concepts, Algorithms, and Applications*. Springer-Verlag, New York, 1996.

[31] J. Foo, X. Wan, and G.E. Karniadakis. The multi-element probabilistic collocation method (ME-PCM): Error analysis and applications. *J. Comput. Phys.*, 227(22):9572–9595, 2008.

[32] B.L. Fox. *Strategies for Quasi-Monte Carlo*. Kluwer Academic, Norwell, Massachusetts, 1999.

[33] P. Frauenfelder, C. Schwab, and R.A. Todor. Finite elements for elliptic problems with stochastic coefficients. *Comput. Methods Appl. Mech. Engrg.*, 194:205–228, 2005.

[34] D. Funaro. *Polynomial Approximation of Differential Equations*. Springer-Verlag, Berlin, 1992.

[35] B. Ganapathysubramanian and N. Zabaras. Sparse grid collocation methods for stochastic natural convection problems. *J. Comput. Phys.*, 225(1):652–685, 2007.

[36] C.W. Gardiner. *Handbook of Stochastic Methods for Physics, Chemistry and the Natural Sciences*, 2nd ed. Springer-Verlag, Berlin, 1985.

[37] S.E. Geneser, R.M. Kirby, D. Xiu, and F.B. Sachse. Stochastic Markovian modeling of electrophysiology of ion channels: Reconstruction of standard deviations in macroscopic currents. *J. Theoret. Biol.*, 245(4):627–637, 2007.

[38] J.E. Gentle. *Random Number Generation and Monte Carlo Methods*. Springer-Verlag, New York, 2003.

[39] T. Gerstner and M. Griebel. Numerical integration using sparse grids. *Numer. Algorithms*, 18:209–232, 1998.

[40] R. Ghanem, S. Masri, M. Pellissetti, and R. Wolfe. Identification and prediction of stochastic dynamical systems in a polynomial chaos basis. *Comput. Methods Appl. Mech. Engrg.*, 194:1641–1654, 2005.

[41] R.G. Ghanem. Scales of fluctuation and the propagation of uncertainty in random porous media. *Water Resources Res.*, 34:2123, 1998.

[42] R.G. Ghanem. Ingredients for a general purpose stochastic finite element formulation. *Comput. Methods Appl. Mech. Engrg.*, 168:19–34, 1999.

[43] R.G. Ghanem. Stochastic finite elements for heterogeneous media with multiple random non-Gaussian properties. *ASCE J. Engrg. Math.*, 125(1):26–40, 1999.

[44] R.G. Ghanem and A. Doostan. On the construction and analysis of stochastic models: Characterization and propagation of the errors associated with limited data. *J. Comput. Phys.*, 217(1):63–81, 2006.

[45] R.G. Ghanem and P. Spanos. *Stochastic Finite Elements: A Spectral Approach*. Springer-Verlag, New York, 1991.

[46] D. Gottlieb and S.A. Orszag. *Numerical Analysis of Spectral Methods: Theory and Applications*. SIAM-CMBS, Philadelphia, Pennsylvania, 1997.

[47] D. Gottlieb and C.-W. Shu. On the Gibbs phenomenon and its resolution. *SIAM Rev.*, 39(4):644–668, 1997.

[48] D. Gottlieb and D. Xiu. Galerkin method for wave equations with uncertain coefficients. *Commun. Comput. Phys.*, 3(2):505–518, 2008.

[49] S. Haber. Numerical evaluation of multiple integrals. *SIAM Rev.*, 12(4):481–526, 1970.

[50] C. Hastings. *Approximations for Digital Computers*. Princeton University Press, Princeton, New Jersey, 1955.

[51] J.S. Hesthaven, S. Gottlieb, and D. Gottlieb. *Spectral Methods for Time-dependent Problems*. Cambridge: Cambridge University Press, 2007.

[52] W. Hörmann, J. Leydold, and G. Derflinger. *Automatic Nonuniform Random Variate Generation*. Springer-Verlag, Berlin, 2004.

[53] T. Hou, W. Luo, B. Rozovskii, and H.M. Zhou. Wiener chaos expansions and numerical solutions of randomly forced equations of fluid mechanics. *J. Comput. Phys.*, 217:687–706, 2006.

[54] R. Kalman and R. Bucy. New results in linear prediction and filter theory. *Trans. ASME J. Basic Engrg.*, 83D:85–108, 1961.

[55] I. Karatzas and S.E. Shreve. *Brownian Motion and Stochastic Calculus*. Springer-Verlag, New York, 1988.

[56] M. Kleiber and T.D. Hien. *The Stochastic Finite Element Method*. John Wiley & Sons, 1992.

[57] P.E. Kloeden and E. Platen. *Numerical Solution of Stochastic Differential Equations*. Springer-Verlag, Berlin, 1999.

[58] O.M. Knio and O.P. Le Maitre. Uncertainty propagation in CFD using polynomial chaos decomposition. *Fluid Dynam. Res.*, 38(9):616–640, 2006.

[59] D.E. Knuth. *The Art of Computer Programming*. Addison-Wesley, Reading, Massachusetts, 1998.

[60] R. Koekoek and R.F. Swarttouw. The Askey-scheme of hypergeometric orthogonal polynomials and its q-analogue. Technical Report 98-17, Department of Technical Mathematics and Informatics, Delft University of Technology, Delft, Netherlands, 1998.

[61] O. Le Maitre, O. Knio, H. Najm, and R. Ghanem. A stochastic projection method for fluid flow: Basic formulation. *J. Comput. Phys.*, 173:481–511, 2001.

[62] O. Le Maitre, O. Knio, H. Najm, and R. Ghanem. Uncertainty propagation using Wiener-Haar expansions. *J. Comput. Phys.*, 197:28–57, 2004.

[63] O. Le Maitre, H. Najm, R. Ghanem, and O. Knio. Multi-resolution analysis of Wiener-type uncertainty propagation schemes. *J. Comput. Phys.*, 197:502–531, 2004.

[64] O. Le Maitre, M. Reagan, H. Najm, R. Ghanem, and O. Knio. A stochastic projection method for fluid flow: Random process. *J. Comput. Phys.*, 181:9–44, 2002.

[65] P. L'Ecuyer. Uniform random number generation. *Handbooks Oper. Res. and Management Sci.* Elsevier, Philadelphia, Pennsylvania, 2004.

[66] J. Li and D. Xiu. On numerical properties of the ensemble Kalman filter for data assimilation. *Comput. Methods Appl. Mech. Engrg.*, 197:3574–3583, 2008.

[67] J. Li and D. Xiu. A generalized polynomial chaos based ensemble Kalman filter with high accuracy. *J. Comput. Phys.*, 228:5454–5469, 2009.

[68] G. Lin, C.-H. Su, and G.E. Karniadakis. Predicting shock dynamics in the presence of uncertainties. *J. Comput. Phys.*, 217(1):260–276, 2006.

[69] G. Lin, C.-H. Su, and G.E. Karniadakis. Random roughness enhances lift in supersonic flow. *Phy. Rev. Lett.*, 99(10):104501-1—104501-4, 2007.

[70] G. Lin, X. Wan, C.-H. Su, and G.E. Karnidakis. Stochastic computational fluid mechanics. *IEEE Comput. Sci. Engrg.*, 9(2):21–29, 2007.

[71] W.K. Liu, T. Belytschko, and A. Mani. Probabilistic finite elements for nonlinear structural dynamics. *Comput. Methods Appl. Mech. Engrg.*, 56:61–81, 1986.

[72] W.K. Liu, T. Belytschko, and A. Mani. Random field finite elements. *Internat. J. Numer. Methods Engrg.*, 23:1831–1845, 1986.

[73] M. Loève. *Probability Theory*, 4th ed. Springer-Verlag, New York, 1977.

[74] W.L. Loh. On Latin hypercube sampling. *Ann. Statist.*, 24(5):2058–2080, 1996.

[75] X. Ma and N. Zabaras. An adaptive hierarchical sparse grid collocation algorithm for the solution of stochastic differential equations. *J. Comput. Phys.*, 228:3084–3113, 2009.

[76] Y. Marzouk and D. Xiu. A stochastic collocation approach to Bayesian inference in inverse problems. *Commun. Comput. Phys.*, 6(4):826–847, 2009.

[77] Y.M. Marzouk, H.N. Najm, and L.A. Rahn. Stochastic spectral methods for efficient Bayesian solution of inverse problems. *J. Comput. Phys.*, 224(2):560–586, 2007.

[78] L. Mathelin and M.Y. Hussaini. A stochastic collocation algorithm for uncertainty analysis. Technical Report NASA/CR-2003-212153, NASA Langley Research Center, Langley, Virginia, 2003.

[79] H. Niederreiter. *Random Number Generation and Quasi-Monte Carlo Methods*. SIAM, Philadelphia, Pennsylvania, 1992.

[80] H. Niederreiter, P. Hellekalek, G. Larcher, and P. Zinterhof. *Monte Carlo and Quasi-Monte Carlo Methods 1996*. Springer-Verlag, Berlin, 1998.

[81] F. Nobile, R. Tempone, and C. Webster. A sparse grid stochastic collocation method for partial differential equations with random input data. *SIAM J. Numer. Anal.*, 46(5):2309–2345, 2008.

[82] A. Nouy, A. Clement, F. Schoefs, and N. Moes. An eXtended stochastic finite element method for solving stochastic partial differential equations on random domains. *Comput. Methods Appl. Mech. Engrg.*, 197:4663–4682, 2008.

[83] E. Novak and K. Ritter. High dimensional integration of smooth functions over cubes. *Numer. Math.*, 75:79–97, 1996.

[84] E. Novak and K. Ritter. Simple cubature formulas with high polynomial exactness. *Constr. Approx.*, 15:499–522, 1999.

[85] B. Oksendal. *Stochastic Differential Equations: An Introduction with Applications*, 5th ed. Springer-Verlag, Berlin, 1998.

[86] S.A. Orszag and L.R. Bissonnette. Dynamical properties of truncated Wiener-Hermite expansions. *Phys. Fluids*, 10:2603–2613, 1967.

[87] B.D. Ripley. *Stochastic Simulation*. John Wiley & Sons, New York, 1987.

[88] M. Rosenblatt. Remark on a multivariate transformation. *Ann. Math. Stat.*, 23(3):470–472, 1953.

[89] A. Sandu, C. Sandu, and M. Ahmadian. Modeling multibody dynamic systems with uncertainties. Part I: Theoretical and computational aspects. *Multibody Syst. Dyn.*, 15(4):369–391, 2006.

[90] C. Sandu, A. Sandu, and M. Ahmadian. Modeling multibody dynamic systems with uncertainties. Part II: Numerical applications. *Multibody Syst. Dyn.*, 15(3):245–266, 2006.

[91] W. Schoutens. *Stochastic Processes and Orthogonal polynomials*. Springer-Verlag, New York, 2000.

[92] C. Schwab and R.A. Todor. Sparse finite elements for elliptic problems with stochastic data. *Numer. Math.*, 95:707–734, 2003.

[93] C. Schwab and R.A. Todor. Karhunen-Loève approximation of random fields by generalized fast multipole methods. *J. Comput. Phys.*, 217:100–122, 2006.

[94] J. Shi and R.G. Ghanem. A stochastic nonlocal model for materials with multiscale behavior. *Internat. J. Multiscale Comput. Engrg.*, 4(4):501–519, 2006.

[95] M. Shinozuka and G. Deodatis. Response variability of stochastic finite element systems. *J. Engrg. Math.*, 114(3):499–519, 1988.

[96] S.A. Smolyak. Quadrature and interpolation formulas for tensor products of certain classes of functions. *Soviet Math. Dokl.*, 4:240–243, 1963.

[97] P. Spanos and R.G. Ghanem. Stochastic finite element expansion for random media. *ASCE J. Engrg. Math.*, 115(5):1035–1053, 1989.

[98] M. Stein. Large sample properties of simulations using Latin hypercube sampling. *Technometrics*, 29(2):143–151, 1987.

[99] A.H. Stroud. *Approximate Calculation of Multiple Integrals*. Prentice-Hall, Englewood Cliffs, New Jersey, 1971.

[100] G. Szegö. *Orthogonal Polynomials*. American Mathematical Society, Providence, Rhode Island, 1939.

[101] A. Tarantola. *Inverse Problem Theory and Methods for Model Parameter Estimation*. SIAM, Philadelphia, Pennsylvania, 2005.

[102] D.M. Tartakovsky and D. Xiu. Stochastic analysis of transport in tubes with rough walls. *J. Comput. Phys.*, 217(1):248–259, 2006.

[103] M.A. Tatang, W.W. Pan, R.G. Prinn, and G.J. McRae. An efficient method for parametric uncertainty analysis of numerical geophysical models. *J. Geophys. Res.*, 102:21925–21932, 1997.

[104] R. Tempone, F. Nobile, and C. Webster. An anisotropic sparse grid stochastic collocation method for elliptic partial differential equations with random input data. *SIAM J. Numer. Anal.*, 46(5):2411–2442, 2008.

[105] A.F. Timan. *Theory of Approximation of Functions of a Real Variable.* Pergamon Press, Oxford, 1963.

[106] J. Todd. *Introduction to the Constructive Theory of Functions.* Birkhäuser Verlag, Basel, 1963.

[107] R.A. Todor and C. Schwab. Convergence rates for sparse chaos approximations of elliptic problems with stochastic coefficients. *IMA J. Numer. Anal.*, 27(2):232–261, 2007.

[108] L.N. Trefethen. Is Gauss quadrature better than Cleanshaw-Curtis? *SIAM Rev.*, 50(1):67–87, 2008.

[109] H.L. Van Trees. *Detection, Estimation and Modulation Theory, Part 1.* John Wiley & Sons, New York, 1968.

[110] X. Wan and G.E. Karniadakis. An adaptive multi-element generalized polynomial chaos method for stochastic differential equations. *J. Comput. Phys.*, 209(2):617–642, 2005.

[111] X. Wan and G.E. Karniadakis. Multi-element generalized polynomial chaos for arbitrary probability measures. *SIAM J. Sci. Comput.*, 28:901–928, 2006.

[112] J. Wang and N. Zabaras. Using Bayesian statistics in the estimation of heat source in radiation. *Int. J. Heat Mass Trans.*, 48:15–29, 2005.

[113] K.F. Warnick and W.C. Chew. Numerical simulation methods for rough surface scattering. *Waves Random Media*, 11(1):R1–R30, 2001.

[114] G.W. Wasilkowski and H. Woźniakowski. Explicit cost bounds of algorithms for multivariate tensor product problems. *J. Complexity*, 11:1–56, 1995.

[115] N. Wiener. The homogeneous chaos. *Amer. J. Math.*, 60:897–936, 1938.

[116] D. Xiu. Efficient collocational approach for parametric uncertainty analysis. *Commun. Comput. Phys.*, 2(2):293–309, 2007.

[117] D. Xiu, R.G. Ghanem, and I.G. Kevrekidis. An equation-free, multiscale approach to uncertainty quantification. *IEEE Comput. Sci. Engrg.*, 7(3):16–23, 2005.

[118] D. Xiu and J.S. Hesthaven. High-order collocation methods for differential equations with random inputs. *SIAM J. Sci. Comput.*, 27(3):1118–1139, 2005.

[119] D. Xiu and G.E. Karniadakis. Modeling uncertainty in steady state diffusion problems via generalized polynomial chaos. *Comput. Methods Appl. Mech. Engrg.*, 191:4927–4948, 2002.

[120] D. Xiu and G.E. Karniadakis. The Wiener-Askey polynomial chaos for stochastic differential equations. *SIAM J. Sci. Comput.*, 24(2):619–644, 2002.

[121] D. Xiu and G.E. Karniadakis. Modeling uncertainty in flow simulations via generalized polynomial chaos. *J. Comput. Phys.*, 187:137–167, 2003.

[122] D. Xiu and G.E. Karniadakis. A new stochastic approach to transient heat conduction modeling with uncertainty. *Int. J. Heat Mass Trans.*, 46:4681–4693, 2003.

[123] D. Xiu and G.E. Karniadakis. Supersensitivity due to uncertain boundary conditions. *Internat. J. Numer. Methods Engrg.*, 61(12):2114–2138, 2004.

[124] D. Xiu and I.G. Kevrekidis. Equation-free, multiscale computation for unsteady random diffusion. *Multiscale Model. Simul.*, 4(3):915–935, 2005.

[125] D. Xiu, D. Lucor, C.-H. Su, and G.E. Karniadakis. Stochastic modeling of flow-structure interactions using generalized polynomial chaos. *J. Fluids Engrg.*, 124:51–59, 2002.

[126] D. Xiu and J. Shen. An efficient spectral method for acoustic scattering from rough surfaces. *Commun. Comput. Phys.*, 2(1):54–72, 2007.

[127] D. Xiu and J. Shen. Efficient stochastic Galerkin methods for random diffusion equations. *J. Comput. Phys.*, 228:266–281, 2009.

[128] D. Xiu and S.J. Sherwin. Parametric uncertainty analysis of pulse wave propagation in a model of human arterial networks. *J. Comput. Phys.*, 226:1385–1407, 2007.

[129] D. Xiu and D.M. Tartakovsky. A two-scale non-perturbative approach to uncertainty analysis of diffusion in random composites. *Multiscale Model. Simul.*, 2(4):662–674, 2004.

[130] D. Xiu and D.M. Tartakovsky. Numerical methods for differential equations in random domains. *SIAM J. Sci. Comput.*, 28(3):1167–1185, 2006.

[131] F. Yamazaki, M. Shinozuka, and G. Dasgupta. Neumann expansion for stochastic finite element analysis. *J. Engrg. Math.*, 114(8):1335–1354, 1988.

Index

aliasing error, 42, 85, 87
Askey scheme, 27–28, 105

best approximation, 30–31, 32
Burgers' equation, 1–2, 4–5, 76–77, 98–99

central limit theorem, 23
Cholesky's decomposition, 46
Clenshaw-Curtis nodes, 82, 86
conditional probability, 19
convergence, 22; almost sure, 22;
    in distribution, 22; Lp, 23;
    in probability, 22; with probability 1, 22
covariance, 17, 21, 47
cubature, 84, 86–87

distribution, 11, 17; continuous, 12; discrete,
    11; inverse of, 15; posterior, 96; prior, 95

expectations, 13–14, 17

Favard's theorem, 26

Gaussian distribution, 12, 14, 18; approximate
    inversion of, 16
generalized polynomial chaos (gPC), 6–7,
    57–59, 64–65; discrete orthogonal
    projection, 83–85; Hermite, 58;
    homogeneous, 65; Jacobi, 58; Legendre,
    58; orthogonal projection, 59, 65;
    strong approximation, 58; weak
    approximation, 60
Gibbs phenomenon, 35–36, 62–64
graded lexicographic order, 66

Hermite polynomials, 29, 106
hypergeometric series, 27, 105

independence, 18
indicator function, 20
integration rule. See cubature; quadrature

Kalman filter, 99–101; ensemble, 99–101
Karhunen-Loeve expansion, 47–48
Kullback-Leibler divergence, 97

Lagrange interpolation, 37, 79–80
Laguerre polynomials, 30, 106–107
Legendre polynomials, 29
lognormal random variable, 60

marginal density, 17
MCS. See Monte Carlo sampling
mean-square convergence, 59
moment equations, 4, 54
moment-generating function, 14–15
moments, 13–14
Monte Carlo sampling, 3, 53–54

normal distribution. See Gaussian distribution

orthogonal polynomials, 25–26, 105–111
orthogonal projection, 31; discrete, 41

perturbation methods, 3, 55–56
probability space, 11
pseudospectral method, 83–85

quadrature, 40

random domain, 89–90
random variables, 10
random vectors, 16–17
Rosenblatt transformation, 46
Runge function, 38
Runge phenomenon. See Runge function

sign function, 36
Smolyak construction, 82
Sobolev space, 33
sparse grid, 82–83, 86
spectral convergence, 34
stochastic process, 20–21;
    second-order stationary, 21;
    strictly stationary, 21
supersensitivity. See Burgers' equation

three-term recurrence, 26, 105

Vandermonde matrix, 37–38, 80

Weierstrass theorem, 30